T0254282

Lecture Notes in Artificial Intelligence 11889

Subseries of Lecture Notes in Computer Science

More information about this series at http://www.springer.com/series/1244

Konstantinos Tserpes · Chiara Renso ·
Stan Matwin (Eds.)

Multiple-Aspect Analysis of Semantic Trajectories

First International Workshop, MASTER 2019
Held in Conjunction with ECML-PKDD 2019
Würzburg, Germany, September 16, 2019
Proceedings

Editors
Konstantinos Tserpes
Harokopio University
Athens, Greece

Chiara Renso
ISTI-CNR
Pisa, Italy

Stan Matwin
Dalhousie University
Halifax, NS, Canada

ISSN 0302-9743 ISSN 1611-3349 (electronic)
Lecture Notes in Artificial Intelligence
ISBN 978-3-030-38080-9 ISBN 978-3-030-38081-6 (eBook)
https://doi.org/10.1007/978-3-030-38081-6

LNCS Sublibrary: SL7 – Artificial Intelligence

This Springer imprint is published by the registered company Springer Nature Switzerland AG
The registered company address is: Gewerbestrasse 11, 6330 Cham, Switzerland

Preface

An ever-increasing number of diverse, real-life applications, ranging from mobile to social media apps and surveillance systems, produce massive amounts of spatio-temporal data representing trajectories of moving objects. The fusion of those trajectories, commonly represented by timestamped location sequence data (e.g. check-ins and GPS traces), with generally available and semantic-rich data resources can result in an enriched set of more comprehensive and semantically significant objects. The analysis of these sets, referred to as "semantic trajectories", can unveil solutions to traditional problems and unlock the challenges for the advent of novel applications and application domains, such as transportation, security, health, environment, and even policy modeling.

Despite the fact that the semantic trajectories concept is not new, we are now witnessing an increasing complexity in the forms and heterogeneity of the enrichment process producing new kinds of trajectory objects. These new objects call for novel methods that can properly take into account the multiple semantic aspects defining this new form of movement data. It is the very nature of the semantic trajectories that makes this analysis challenging. For instance, the data sources and formats are largely heterogeneous, placing hurdles in the fusion process; or their volumes are too large to process them in conventional ways. In the other cases the state of the semantic trajectories is updated at such a rapid pace, that it is very hard to explore them so as to get an indication of their latent semantics, or even process them in a consistent way since they cannot be stored. Another typical problem is with their unreliable and erroneous nature, where signals are arriving in a mixed order, with gaps and even errors. Similarly, the multiple aspects nature of semantic trajectories increases the difficulty of trajectory pattern mining.

The MASTER 2019 workshop was held in Würzburg, Germany, on September 16, 2019, in conjunction with ECML/PKDD 2019. The format of the workshop included a keynote speech and eight technical presentations. The workshop was attended by around 20 people on average.

This year we received 12 manuscript for consideration, from authors based in 8 distinct countries, from Japan, to Europe, to Brazil, and Canada. After an accurate and thorough single-blind review process with the help of the 22 members of the Program Committee, we selected 8 full papers for presentation at the workshop. The review process focused on the quality of the papers, their scientific novelty and applicability to existing Semantic Trajectory Analysis problems and frameworks. The acceptance of the papers was the result of the reviewers' discussion and agreement. All the high-quality papers were accepted, and the acceptance rate was 66.66%. The accepted articles represent an interesting mix of techniques to solve recurrent as well as new problems in the Semantic Trajectory domain, such as data represetnation models, data management systems, machine learning approaches for anomaly detection, and common pathways identification.

The workshop program was completed by the invited talk entitled "Learning from our movements – The mobility data analytics pipeline" by Prof. Yannis Theodoridis from the University of Piraeus, Greece.

We would like to thank the MASTER 2019 Program Committee, whose members made the workshop possible with their rigorous and timely review process. We would also like to thank ECML/PKDD for selecting and hosting the workshop. Most importantly we would like to thank the emerging community of the Semantic Trajectories' domain that attended the workshop from practically all around the world.

The workshop has been supported by the MASTER project (http://www.master-project-h2020.eu), which has received funding from the European Union's Horizon 2020 research and innovation program under the Marie Skłodowska-Curie grant agreement No 777695.

October 2019

Konstantinos Tserpes
Chiara Renso
Stan Matwin

Organization

Program Committee

Gennady Andrienko	Fraunhofer, Germany
Maria Luisa Damiani	University of Milan, Italy
Magdalini Eirinaki	San Jose State University, USA
Angelo Furno	Université de Lyon, France
Sebastien Gambs	Université du Québec à Montréal, Canada
Ralf Hartmut Güting	Fernuniversität Hagen, Germany
Sergio Ilarri	University of Zaragoza, Spain
Dimitris Kotzinos	University of Cergy-Pontoise, France
Jose Macedo	Federal University of Ceara, Brazil
Stan Matwin	Dalhousie University, Canada
Dimitrios Michail	Harokopio University, Greece
Anna Monreale	University of Pisa, Italy
Mirco Nanni	ISTI-CNR, Italy
Latifa Oukhellou	IFSTTAR, France
Cyril Ray	École Navale, France
Chiara Renso	ISTI-CNR, Italy
Matthias Renz	Christian-Albrechts-Universität zu Kiel, Germany
Cyrus Shahabi	University of Southern California, USA
Amilcar Soares	Institute for Big Data Analytics, Canada
Luis Torgo	University of Porto, Portugal
Goce Trajcevski	Iowa State University, USA
Konstantinos Tserpes	Harokopio University, Greece
Fabio Valdés	FernUniversität in Hagen, Germany
Iraklis Varlamis	Harokopio University, Greece
Demetrios Zeinalipour-Yazti	University of Cyprus, Cyprus

Contents

Learning from Our Movements – The Mobility Data Analytics Era

Yannis Theodoridis(✉)

Data Science Laboratory, University of Piraeus, Piraeus, Greece
ytheod@unipi.gr
http://www.datastories.org/

Abstract. From the pioneering works on spatiotemporal databases back in '90s to the era of Big Mobility Data Analytics nowadays, in this paper we try to follow the thread of research objectives and initiatives in the field. Initially, we provide a flashback to the 25 past years (though from a personal, hence, biased point of view). Then, we discuss in brief the challenges related to mobility data processing, management, analytics, and visualization to be addressed in modern applications tracking populations of moving objects in real time.

1 Introduction

Once upon a time, it was the ChoroChronos EU research project. The challenge at that time ('90s) was to bring spatial and temporal database aspects together in a, then emerging, integrated spatio-termporal domain. As time was passing, new challenges appeared and addressed by the researchers of the field: efficient system architectures, knowledge discovery from mobility data, privacy aspects, etc. Nowadays, in the era of Data Science and Big Data, mobility data analytics aims at learning from objects' movements, covering a range of methods and solutions, from de-noising of location information and integrating with multiple

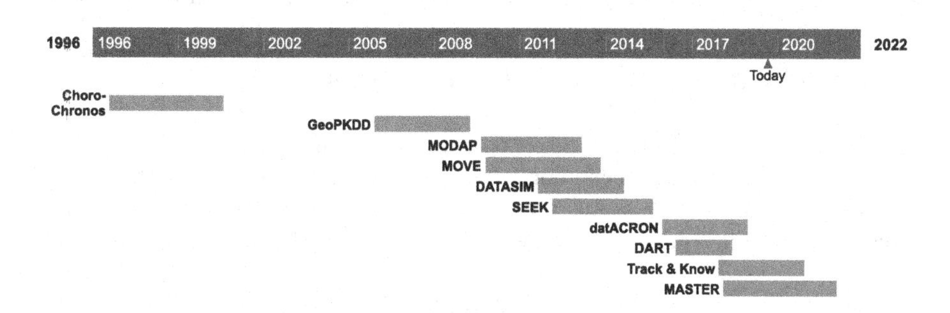

Fig. 1. Related research projects timeline (mid '90s - today).

K. Tserpes et al. (Eds.): MASTER 2019, LNAI 11889, pp. 1–5, 2020.
https://doi.org/10.1007/978-3-030-38081-6_1

heterogeneous related sources to predictive analytics, offline and online. Figure 1 illustrates a timeline of related research projects covering the period from mid '90s until today[1].

2 Flashback to the Past

Let's catch the thread from the beginning... Back in '90s, research in spatial and temporal databases, separately, resulted in pretty mature results to contribute in real-world DBMSs. In the spatial database field, Oracle introduced Spatial Data Option in 1996, OGC released its first specifications in 1997, and PostGIS was launched in 2001. On the other hand, TSQL2 language specification was developed in 1993. The "marriage" of the two fields was led by research projects in both US and Europe; focusing on the latter, the notable ChoroChronos EU project (1996–2000) aimed at bringing together the two communities and integrate their ideas in the so-called **spatio-temporal databases**, where time would be considered a first-class citizen [2,4,5].

What followed was the focus of research in point objects due to the popularity of related applications (tracking of moving objects via GPS technology), which led to the "**moving object trajectory**" concept[2] and, as expected, raised challenges on knowledge discovery from this new type of data as well as on personal data privacy. This brought the "dialogue" with other than database management domains, including machine learning/data mining and data privacy and security. For instance, the GeoPKDD (2005–2009), MODAP (2009–2012) and MOVE (2009–2013) EU projects aimed at **devising knowledge discovery and (privacy-preserving) analysis methods for trajectories of moving objects**, bringing together ICT researchers and domain specialists [3,8]. The advances in social networks and linked open data in '00s also resulted in the so-called location-based social networking and relevant applications. Thus, a new trajectory variation was born, the **semantic trajectories** [6], studied by e.g. the SEEK EU project (2012–2015), the objective of which was to envisage a new semantic enriched knowledge discovery process where the semantic aspect (in the sense of the meaning of the movement) would be embedded in each step.

3 Nowadays - Mobility Data Analytics

Nowadays, one can find plenty of sensor data and open sources of related information, mature Big Data technologies, plethora of Data Science methods and tools. In this environment, a hot research topic is that of **Mobility Data Analytics** (MDA). The range of processes covered under this term includes data

[1] Disclaimer: the flashback in past is biased since it only refers by name to projects where the author has participated. The author's intention is by no means to provide an exhaustive survey of research activities related to the topic of the article.

[2] According to DBLP (dblp.uni-trier.de), the first papers with this term in their title appeared in 2000.

acquisition and processing (typically from multiple and varying data sources), data management (storage and indexing, as usually...), data mining, data privacy, data visualization and user interaction [7].

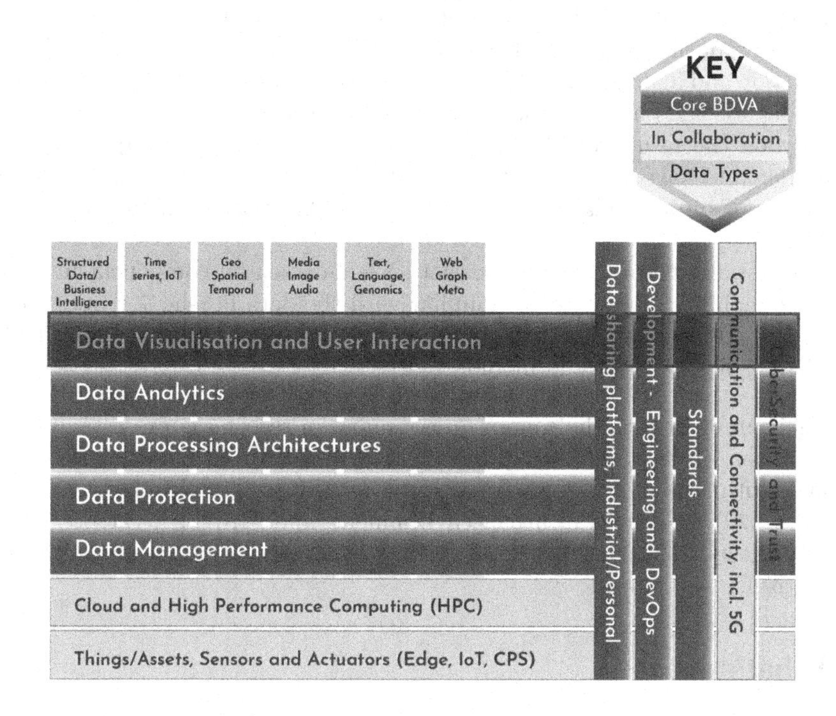

Fig. 2. The BDVA reference model (source: http://www.bdva.eu).

A typical MDA application tracks in real-time a population of humans / vehicles / vessels / aircrafts and handles the resulted trajectories, enriched with heterogeneous context, in order, for instance, to be able to assess traffic situation or drivers' behavior, forecast anticipated movements in short-term or schedule traffic in long-term, react as soon as an "anomaly" is detected, and so on [9]. In this framework, challenges touch almost each architectural layer of an MDA system, compatible with the EU BDVA reference model (illustrated in Fig. 2):

- **data sources** of interest include, on the one hand, streaming information, such as the GPS signals transmitted by the objects themselves, the objects' locations tracked by an external device (e.g. radar), and live weather information, and, on the other hand, archive collections of census, meteorological, etc. data; all of this information to be correctly and smoothly integrated;
- **data processing** requires cleansing (de-noising, smoothing) and semantic enrichment of incoming data as already mentioned, segmentation into trajectories that make sense (e.g. from/to predefined places of interest), and storage in appropriate (relational or NoSQL) stores;

- **data management** includes efficient querying and retrieval, which assume implementation of query algorithms and maintenance of indexing mechanisms suitable for these purposes;
- **data analytics** requires both offline and online algorithms: cluster and frequent pattern analysis for detecting typical movement patterns as well as interesting outliers can run offline on the historical data whereas movement prediction and anomaly detection over the incoming stream should run online; in order for the online methods to be effective, they should take into consideration the results of the offline methods (a now tracked object may be considered to have anomalous behavior if it is quite dissimilar to past typical patterns or quite similar to past outliers) and this makes things even more challenging;
- **data visualization** and user interaction is an essential component in order for the mobility analyst to get familiar with the data he/she is requested to analyze and interact with the above methods and tools; visual analytics (VA) is challenging, especially in mobility data [1].

This is, more or less, the context of a number of recent EU projects, including datAcron (http://datacron-project.eu; 2016–2018), where the use cases are on maritime and aviation, Track & Know (https://trackandknowproject.eu; 2018–2020) on drivers in urban environment, and MASTER (http://www.master-project-h2020.eu; 2018–2022) on land transportation, sea monitoring, and tourism.

4 What's Next

Fortunately, research is an everlasting story. So, what's next? In the near future, we expect to see advances, for instance, in self-organizing and self-cleansing information integration tools (do we know what information is relevant to enrich a location or assess its accuracy better than a crawler searching the Semantic Web?), close to storage-less architectures following the IoT paradigm (if data resides in its original source, what is the need to replicate it in our storage? what if data is unaffordable huge and only meta-data as well as patterns are stored locally?), and Explainable AI (why this and not that technique in order to analyze location data? our right to explanation). We expect these and other even more exciting research outcomes to appear in the years to come.

References

1. Andrienko, G.L., Andrienko, N.V., Bak, P., Keim, D.A., Wrobel, S.: Visual Analytics of Movement. Springer, Berlin (2013). https://doi.org/10.1007/978-3-540-75177-9_14
2. Frank, A.U., et al.: Chorochronos: a research network for spatiotemporal database systems. SIGMOD Rec. **28**(3), 12–21 (1999)

3. Giannotti, F., Pedreschi, D. (eds.): Mobility, Data Mining and Privacy - Geographic Knowledge Discovery. Springer, Berlin (2008). https://doi.org/10.1007/978-3-540-75177-9
4. Güting, R.H., et al.: A foundation for representing and querying moving objects. ACM Trans. Database Syst. **25**(1), 1–42 (2000)
5. Koubarakis, M., et al.: Spatio-Temporal Databases. LNCS, vol. 2520. Springer, Heidelberg (2003). https://doi.org/10.1007/b83622
6. Parent, C., et al.: Semantic trajectories modeling and analysis. ACM Comput. Surv. **45**(4), 42:1–42:32 (2013)
7. Pelekis, N., Theodoridis, Y.: Mobility Data Management and Exploration. Springer, New York (2014). https://doi.org/10.1007/978-1-4939-0392-4
8. Renso, C., Spaccapietra, S., Zimányi, E. (eds.): Mobility Data: Modeling Management and Understanding. Cambridge University Press, Cambridge (2013)
9. Vouros, G.A., et al.: Big data analytics for time critical mobility forecasting: recent progress and research challenges. In: EDBT, pp. 612–623 (2018) OpenProceedings.org

Uncovering Hidden Concepts from AIS Data: A Network Abstraction of Maritime Traffic for Anomaly Detection

Ioannis Kontopoulos(✉), Iraklis Varlamis(✉), and Konstantinos Tserpes(✉)

Department of Informatics and Telematics, Harokopio University of Athens, Athens, Greece
{kontopoulos,varlamis,tserpes}@hua.gr
http://www.dit.hua.gr/

Abstract. The compulsory use of Automatic Identification System (AIS) for many vessel types, which has been enforced by naval regulations, has opened new opportunities for maritime surveillance. AIS transponders are rich sources of information that everyone can collect using an RF receiver and provide real-time information about vessels' position. Properly taking advantage of AIS data, can uncover potential illegal behavior, offer real-time alerts and notify the authorities for any kind of anomalous vessel behavior. In this article, we extend an existing network abstraction of maritime traffic, that is based on nodes (called way-points) that correspond to naval areas of long stays or major turns for vessels (e.g. ports, capes, offshore platforms etc.) and edges (called traversals) that correspond to the routes followed by vessels between two consecutive way-points. The current work, focuses on the connections of this network abstraction and enriches them with semantic information about the different ways that vessels employ when traversing an edge. For achieving this, it proposes an alternative of the popular density based clustering algorithm DB-Scan, which modifies the proximity parameter (i.e. *epsilon*) of the algorithm. The proposed alternative employs in tandem the difference in (i) speed, (ii) course and (iii) position for defining the distance between two consecutive vessel positions (two consecutive AIS signals received from the same vessel). The results show that this combination performs significantly better than using only the spatial distance and, more importantly, results in clusters that have very interesting properties. The enriched network model can be processed and further examined with data mining techniques, even in an unsupervised manner, in order to identify anomalies in vessels' trajectories. Experimental results on a real dataset show the network's potential for detecting trajectory outliers and uncovering deviations on a vessel's route.

Keywords: Trajectory analytics · AIS vessel monitoring · Anomaly detection

K. Tserpes et al. (Eds.): MASTER 2019, LNAI 11889, pp. 6–20, 2020.
https://doi.org/10.1007/978-3-030-38081-6_2

1 Introduction

Today's maritime surveillance systems are constantly flooded by data coming from AIS transponders, which are embedded in vessels. The use of AIS transponders was made compulsory for all vessels over 300 Gross Tonnage and all passenger vessels in 2002 by the Regulation 19 of SOLAS Chapter V[1]. However, even smaller vessels, from yachts to fishing boats [1], are now using AIS to report their positions to the nearby vessels, usually for safety purposes, making AIS the number one system for global vessel tracking. Each vessel transmits two kinds of AIS data, dynamic and static. The former, periodically sends data regarding vessel's position, speed and heading. The transmission rate depends on the vessel's speed and becomes higher when the speed is greater. The latter, sends data, every six minutes approximately, regarding vessel's destination, type, size and draught of its hull.

Due to the fact that AIS data are sent periodically with high transmission rates, they are of utmost importance to the maritime authorities for vessel tracking purposes. Therefore, a system that takes advantage of such data and is able to notify the authorities in real-time for any abnormal vessel behavior can be valuable for the authorities. This work contributes directly towards anomaly detection from AIS data. It builds upon our previous work in the field [2], which defined a methodology for extracting a network abstraction of the maritime traffic in an area. The input in that work was a lengthy log of AIS data collected from vessels that sailed in that area and the output was a network representation model of the typical routes that the vessels have followed. In that network representation, the nodes (also called *way-points*) are regions of special interest for the routes of vessel and they usually correspond to ports, capes, offshore platforms etc., where multiple vessels usually stop for short or longer periods, or perform major changes in their direction. Similarly, the connections between nodes represent the vessel movement from one way-point to another and thus, a vessel trajectory is a traversal of the network, from a certain way-point to a distant way-point. This traversal either follows the existing connections (and the trajectory can be considered normal) or deviates and hops from one node to another, not directly connected, node. The aggregated information from all vessels that crossed a network connection are used to extract features for this connection (a potential sub-trajectory for other vessels), such as the average, minimum and maximum speed etc.

In this work, we take this simple aggregation one step ahead, and provide a methodology that can be used to process this multi-vessel information in a more proficient manner. The proposed method adds richer information to each connection that have been traversed by multiple vessels. To extend the previously proposed network abstraction we use a clustering algorithm, that manages to identify different movement patterns for the same connection. This information is then used as a reference in the analysis of a vessels' journey and can allow to identify routes that deviate from the previously extracted patterns. Furthermore,

[1] http://solasv.mcga.gov.uk/regulations/regulation19.htm.

it builds upon the semantic information of the edges of the network abstraction and adds to these connections common patterns the vessels must follow in order to travel between two way-points. Therefore, the common pathways and behavior of the vessels in terms of space, speed and heading are integrated to the already proposed network abstraction.

The main idea behind this work is that vessels of the same type (e.g., cargo vessels) that travel towards the same destination, follow common routes that pass through certain way-points and have similar moving patterns such as the same speed or heading. The major contributions of this work are:

- A variation of the popular density based clustering algorithm (DBScan) that takes into account the difference in speed and course as well as the spatial distance of trajectory points and extracts common navigation behaviors.
- A framework for taking advantage of these common navigation behaviors, by constructing movement models for different regions and vessel types and using them to detect deviations from the models.

A framework like this, allows further analysis by using well-known network analysis or data mining techniques enabling easier understanding of the maritime traffic.

The rest of the paper is structured as follows. Section 2 summarizes the literature in the field of feature extraction from multiple trajectories and their use for trajectory comparison. It focuses on works that summarize historical data and build semantic models for an area. In Sect. 3, the proposed methodology of enriching the network abstraction model is presented in detail and Sect. 4 discusses the preliminary results of our methodology in anomaly detection. Finally, Sect. 5 concludes the paper by summarizing the presented methodology and highlighting the impact of this work in the domain of the maritime surveillance by showing the possible use cases in the field of anomaly detection.

2 Related Work

In the context of the proposed work, traffic network abstraction and anomaly detection is the main focus. As a network abstraction model, it is comparable to methodologies that compress or summarize trajectories from historical AIS data in order to improve maritime surveillance systems. As a methodology for anomaly detection it is comparable to techniques that use historical AIS data to detect abnormal or noteworthy patterns or events.

Several works on maritime surveillance have used grid partitioning of the surveillance area into tiles or hexagons [3] for mapping vessel trajectories to polylines or sequences of spatial indexes or key-points [4]. The proposed model is a more coarse-grained representation than other trajectory simplification methods that try to remove redundant AIS data, but still keeping a large amount of them. Such methods apply to single vessel trajectories, whereas the proposed method applies to multiple vessel trajectories in the same region. The proposed methodology results with a few key-points extracted from the set of trajectories – the way-points – and a set of edges between them, that contain statistics extracted from the actual vessel trajectories, which are clustered by similarity. Section 3 shows that the edges connect way-points that are away from each other and edges contain sufficient information about the vessels' journeys between each pair of way-points.

Many works the recent years try to build maritime traffic network representations from historical AIS data [5,6]. Arguedas et al. [5] propose a two-layer network: (i) an external layer that uses way-points as nodes/vertices and routes as edges/lines and (ii) an internal layer that consists of nodes or *breakpoints* that represent vessels' changes in behavior and edges or *tracklets* that represent vessel trajectories. The former layer is a traffic network abstraction, while the latter is a network that provides information about each vessel layer individually. While an edge in the first layer can a be a route from a port to another port, an edge in the internal layer comprises all the simplified trajectories (using Douglas-Peucker algorithm [7]) that sailed across this route.

The complexity of the internal layer raises scalability issues that can be seen in the analysis of a real dataset. It is characteristic that the use of the 454 complete port-to-port routes in the small area of the Baltic sea resulted in an internal layer with 2,095 tracklets. Our proposed model is similar to that of the external layer of [5] but provides a much richer internal layer, that maintains statistical information extracted from the trajectories of the sailing vessels. The resulting model significantly reduces the total amount of data contributed originally by the vessels, without loosing its descriptive power.

Since maritime traffic networks are able to provide compressed information about vessel trajectories, their use seems to be essential for vessel motion analysis and abnormal behavior. The problem of anomaly detection in the maritime domain [8] has been the focus of research for many years, although in the recent years it started attracting more attention. From the early works on anomaly detection from Holst et al. [9] and the later works of Varlamis et al. [10] and Chatzikokolakis et al. [11] on the detection of search and rescue patterns, several representation models and algorithms have been developed to increase maritime situation awareness, identify potential illegal activities and detect anomalous patterns in the vessels' trajectories.

In [18], Pallota et al. propose a methodology for anomaly detection through the use of a maritime traffic model. The model first extracts way-points or clusters from vessel positions or ports and creates or updates the properties of the vessels in the surveillance area. Way-points are extracted and route objects are created by clustering the extracted vessel flows, using the DB-Scan algorithm, which contain spatio-temporal and kinematic features. Probabilities are extracted to classify a set of vessel positioning observations to a route, then using the classified route a prediction is made for the future location. Finally, transition probabilities are used to detect if a vessel's behavior deviates from normality. Authors in [12] compare two methodologies for anomaly detection which both use the Gaussian Mixture Model (GMM) with a different algorithm for clustering. The first one uses the Expectation Maximization (EM) algorithm while the second one uses the greedy version of the EM algorithm. Both techniques consider momentary states of the vessel motion. As an extension of the approach proposed in [12], authors in [13] evaluate two models for detecting anomalies and their ability to distinguish simulated trajectories from real ones, the GMM and the Kernel Density Estimator (KDE). Results indicated that there is no significant difference in the performance of these two models.

The proposed solution is expected to perform better than related frameworks for anomaly detection from AIS data, which employ the position information of the consecutive vessel signals that constitute its trajectory and use Euclidean or other distance metrics in a two-dimensional space (i.e., latitude and longitude) [14,15] or probabilistic approaches that partition space into tiles and estimate the probability of vessels to appear in a certain sequence of tiles [13] ignoring speed and direction. Even in approaches that use historical data to extract the average speed [16] or direction of move in a certain area [18], or techniques such as Piecewise Linear Segmentation (PLS) [17], speed and direction information are used only for predicting future vessel position and the detection of deviation always measures the spatial distance of the actual from the predicted position. From our knowledge, this is the first approach that builds a composite model of speed, direction and position for trajectories, which is then used to directly detect deviations of any of the three features or any combination of them. It is also expected to provide a richer model for the comparison of whole trajectories or sub-trajectories than the techniques that employ equal length sub-trajectories, or dynamic time warping and spatial distances to compare trajectories [19,20] or techniques that combine spatial and temporal dimensions for indexing trajectories [21].

3 The Proposed Approach

The proposed approach is applied to AIS data collected from multiple vessels of the same type (e.g., cargo vessels) for a predefined period of time and a predefined bounding box (e.g., geographic surveillance area of interest), but it is also applicable to larger geographic areas, periods of time and more types of vessels. Since, different types of vessels vary in size and shape, they may follow different routes even if they want to reach the same destination. Furthermore, specific vessel types such as cargo vessels might make much more intermediate stops (e.g. in middle sea platforms) than others. Although, the network abstraction model is the same for all types of vessels, the detailed information that it carries may vary per vessel type. So, in the following we present the model and the way its information is extracted but we demonstrate our approach on an AIS dataset from cargo vessels only.

The main steps of our approach are illustrated in Fig. 1.

- In the *route identification* step, the way-points are extracted from multi-vessel trajectory data, following a methodology proposed in [2] and summarized in Sect. 3.1. Vessel trajectories are then expressed as sequences of sub-trajectories that connect intermediate way-points.

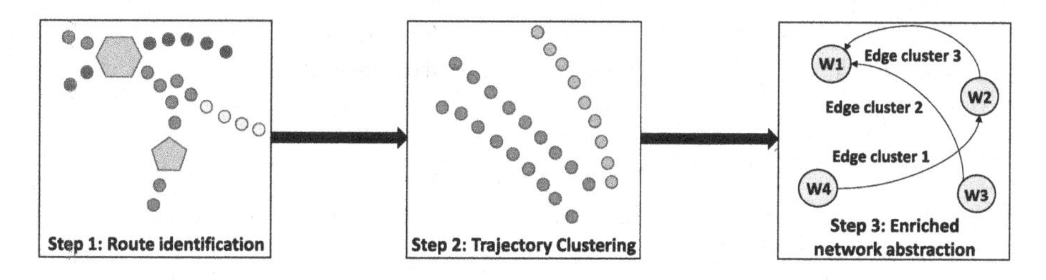

Fig. 1. The steps of the proposed approach

- The (sub-)*trajectory clustering* step is the main contribution of this work, which introduces a novel use of the DB-Scan algorithm that takes into account 3 parameters to identify neighboring points. The methodology followed in this step is explained in details in Sect. 3.2.
- In the *network abstraction model enrichment* step, several statistics are extracted for each cluster. The statistics summarize the movement of multiple vessels along the network edge. The details of these statistics and their extraction method is given in Sect. 3.3.

The final output model, comprises a set of way-points (vertices) dispersed across the monitored region and several sub-trajectory clusters (edges) with their statistics per cluster to represent the different ways of moving between two way-points. This output can be used for many use cases in the field of anomaly detection.

3.1 Route Identification

The first step of our methodology is the identification of way-points, which represent areas where many vessels have stopped (stop points) or did a major directional change (turn points) in the past. As already demonstrated in [2], way-points are created by clustering stop and turn points using a spatial density clustering algorithm (i.e. DB-Scan). The resulting way-points are the nodes of the network abstraction model, which contains information about way-points' size and density (number of stop or turn points per area unit). The size and density of way-points is strongly connected to the parameters of the DB-Scan algorithm. In our working examples, we focus only on the bigger way-points (i.e. those that contain more than 50 points). The idea behind this filtering is that bigger and denser way-points would belong to the trajectories of more vessels.

In our prototype analysis, we focus only on the trajectories that have at least 2 way-points, although the same methodology can be applied in all trajectories and respectively to all the edges of the network. Using different selection thresholds may result either in losing semantic information or in keeping too much information and this is a subject of further experimentation. For example, using higher thresholds (e.g. keeping even larger way-points only) will result in a higher level of abstraction and will probably loose the fine grained details of multiple vessel patterns, whereas using lower thresholds will result in keeping too much information and achieve low or no abstraction at all.

3.2 Trajectory Clustering

The second step refers to the clustering of the trajectories that have the same origin and destination way-points. The typical algorithm for clustering the points of one or more trajectories is DB-Scan [22], which is employed as a density-based spatial clustering method. DB-Scan takes two parameters, *epsilon* which specifies how close two points must be to be considered neighbors, and *minPts* which specifies the number of neighbors a point must have to be included in a cluster. Our proposed DB-Scan version uses 3 parameters to specify the proximity of candidate vessel AIS signals (positions):

- s: absolute difference of the speed between two positions (speed-based)
- h: absolute difference of the course over ground between two positions (heading-based)
- *eps*: harvesine distance between two positions (spatial-based)

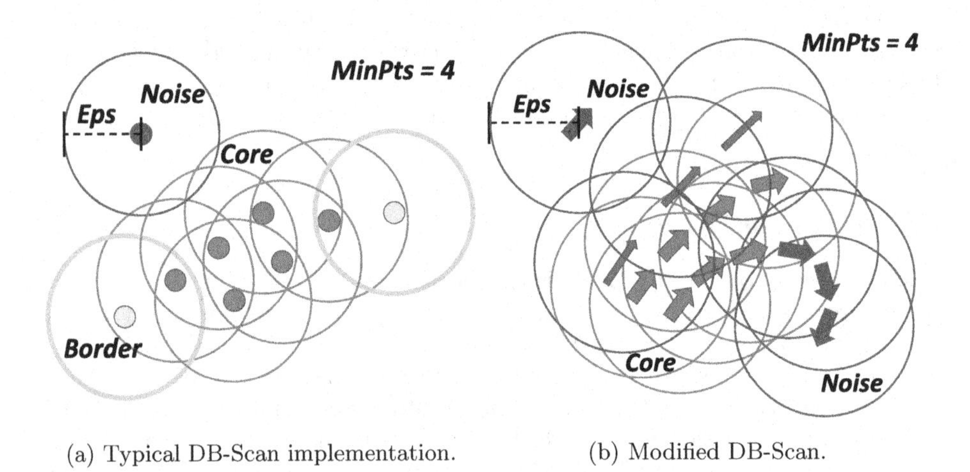

(a) Typical DB-Scan implementation. (b) Modified DB-Scan.

Fig. 2. Comparison of DB-Scan implementations.

To the best of our knowledge, this DB-Scan variation has not been used in the related literature. Therefore, each vessel position contains three types of information: (i) the vessel speed at this position, (ii) the vessel course over ground at this position, (iii) the latitude and longitude of the position. Also, for a vessel position to be clustered together with another vessel position, the absolute difference in their speed must be below a threshold s, the absolute difference in their heading below a threshold h and the distance between them must be below a threshold *eps* at the same time. This type of clustering groups together trajectory points that have similar speed, heading and are close to each other. An example of this type of clustering can be seen in Fig. 2 which compares the two implementations of the DB-Scan algorithm. Figure 2a shows the typical DB-Scan implementation, which creates a cluster if points are spatially close to each other. On the other hand, Fig. 2b illustrates the modified DB-Scan for the positions of moving objects, which considers two points (actually two vectors with position, direction and speed) to be in the same neighborhood when the vectors' positions are spatially close to each other, but they also have similar direction and speed. In the modified version blue arrows indicate noise vectors, which are either away, or have different speed or have a different direction from all their neighbouring vectors.

To have more accurate clustering results, we exclude positions that are located inside the way-points. Since way-points are areas of interest through which vessels frequently pass, it can be easily inferred that the way-points might be ports, platforms, canals or waterways. Inside these way-points, vessels tend to alter their speed or heading frequently, which may corrupt the clustering results.

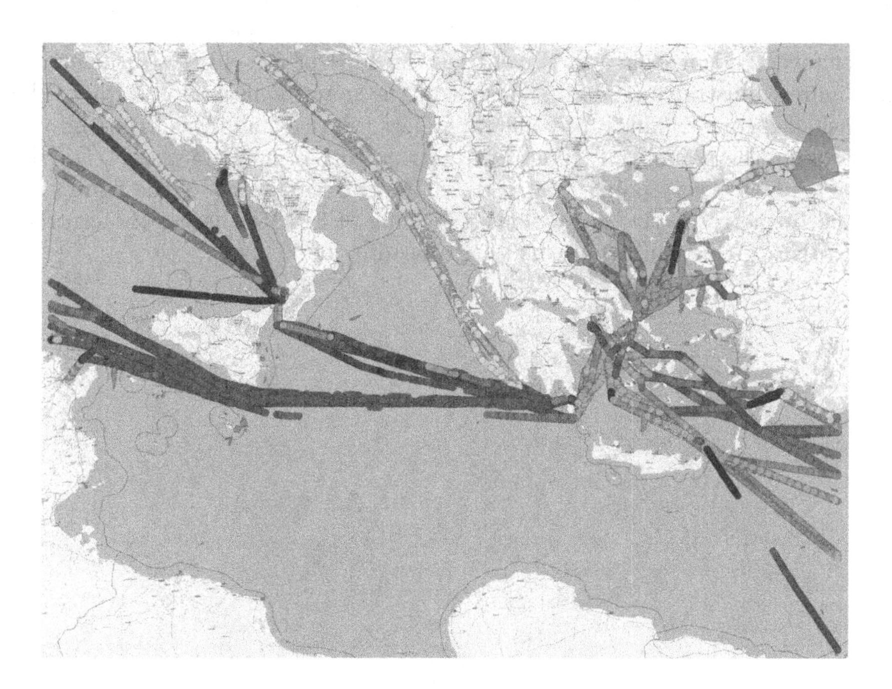

Fig. 3. Example of the trajectory clustering.

Figure 3 illustrates the result of running the proposed trajectory clustering method[2] to all cargo vessels that sail in the east Mediterranean sea and are headed to the port of Piraeus, Greece, using $s = 3$, $h = 3$, $eps = 20\,\text{km}$ and

[2] DB-Scan parameters have been empirically determined.

$minPts = 10$. We can see that trajectories with similar speed and heading are placed in the same cluster, which resembles to the behavior of the basic DBScan (e.g. the cluster formed in the Adriatic sea), whereas points of the same trajectory may belong to different clusters, even though they are spatially close, because of the differences in speed or heading (e.g. the clusters that are formed near the port of Tripolis, Lybia, on the left part of Fig. 3).

3.3 Enriched Network Abstraction

The final step of the process is the enrichment of the network model with information about the clusters of sub-trajectories in each network edge (or in selected edges, e.g. the most frequently traversed). Since, we have created clusters of trajectories (edges) between way-points, we can add information to these edges to form a comprehensive network of the maritime traffic. To this end, for each cluster or edge of the network we calculate the average travelling speed and heading of the vessels. Moreover, the typical deviation of these values is also calculated. Finally, the start point and the end point of the cluster are computed (beginning and ending of the trajectories) along with the average temporal distance of each cluster (average time taken to travel from the start of the cluster to the end of it). Figure 4 illustrates a small snapshot of the network near Sicily, Italy. The green shaded convex hulls represent way-points (vertices) and the green and yellow dots are the points that comprise the trajectory of a single vessel[3]. For the (yellow) subtrajectory points that connect the two way-points of the figure, the centroid of the respective cluster has a heading of 319.15 whereas the centroid of the other (green) (subtrajectory) has a heading of 322.28.

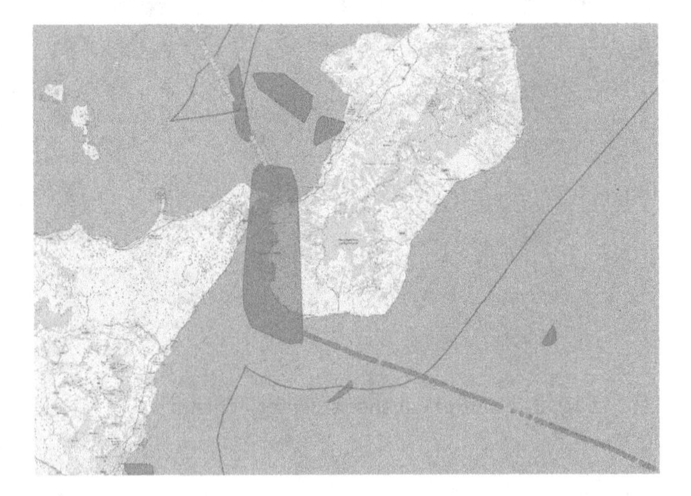

Fig. 4. Example of the edges of the network abstraction. (Color figure online)

[3] For demonstration purposes, this cluster contains points from a single vessel's trajectory.

4 Application to a Real Dataset

To examine the results of our enriched network model, a dataset provided by MarineTraffic[4] was used, containing 2.9 million AIS messages received from 1,716 distinct "cargo" vessels sailing in the eastern half of the Mediterranean Sea during August 2015. Since no information about the existence of anomalous behaviors existed in this dataset, we employed unsupervised techniques to detect potential anomalies or outliers. Although outliers can be detected, further examination is required to understand the reason behind the unusual behavior and the characteristics of the trajectories selected.

4.1 Network Creation from Real AIS Positions

The first step in building the enriched network abstraction is the creation of the way-points (vertices of the network). The identification of the way-points is a two-step process that requires to *i*) identify key-points in the trajectories of the vessels and *ii*) spatially cluster together dense key-points. To identify the key-points we used a speed threshold of 2 knots and a bearing rate threshold of 0.1 degrees per minute, which resulted in several thousand low speed AIS positions and turns in the trajectories of the surveillance area. To create the clusters of key-points, we used the DB-Scan algorithm with a minimum number of ten key-points ($minPts = 10$) within a radius of 2 km ($eps = 2000$), resulting in 616 clusters.

The second step involves clustering of the trajectories with similar characteristics. For this step, we grouped the trajectories per destination and applied the proposed modified version of DB-Scan, which requires for 3 parameters to be satisfied in order for a point to be considered a neighboring one (speed-based, heading-based, spatial-based). For a point to be in the same cluster, its speed must not differ more than 3 knots and its heading more than 3° within a 20 km radius. Moreover, a minimum number of 10 points is required to form a cluster.

In the remainder of this section we demonstrate cases of vessels that had unusual behavior in terms of the way they deviate from their route or in terms of the way they suddenly change course to reach the same destination.

4.2 Detection of Outliers in the Trajectories

The lack of Maritime Situational Awareness (MSA) is a key factor in many incidents that are due to crew fatigue, stress or even engine failures, despite the major improvements in maritime safety. A sudden change in the course of a vessel is considered a noteworthy or anomalous event for the maritime authorities for several reasons, either due to human factors or technical ones. Several cases have been recorded in the past, in which engines fail during a vessel's voyage and the vessel starts drifting away from its normal route. This type of deviation in a vessel's route could potentially lead to collisions with nearby vessels

[4] https://www.marinetraffic.com/.

or collisions with rocky islands, endangering multiple vessels in the vicinity or the environment (e.g., oil spills). Such small deviations from the normal route cannot be detected by algorithms that seek for major turns, and the same holds for temporal decelerations or accelerations and algorithms that seek for sudden stops. Similarly, when vessels are in distress due to piracy attacks or when they take part in search and rescue operations and they perform manoeuvres, it is not always feasible to detect such combined actions that include speed and route change and deviation from the normal route. These types of behavior require an immediate course of action by the authorities. The proposed network abstraction model, with the information it carries on each edge concerning the clusters of movement patterns (in terms of speed, course over ground and location) is able to capture such cases that comprise small or larger deviations in the trajectories. A few outlier cases that have been detected (Fig. 5) on a real dataset are presented in the following.

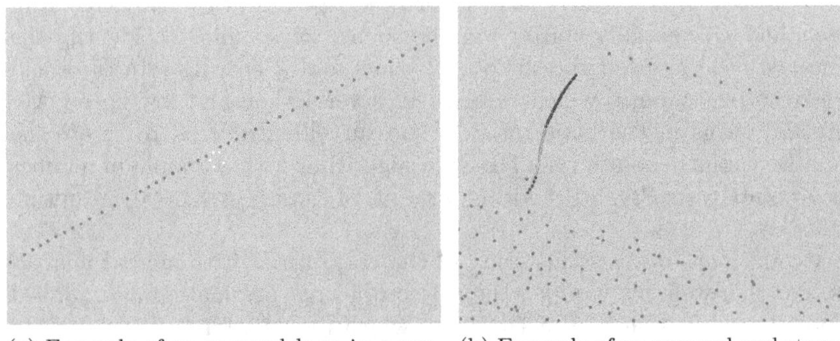

(a) Example of an unusual loop in a vessel's trajectory.

(b) Example of an unusual and steep deviation of a vessel.

(c) A trajectory that does not follow the usual maritime traffic has been detected.

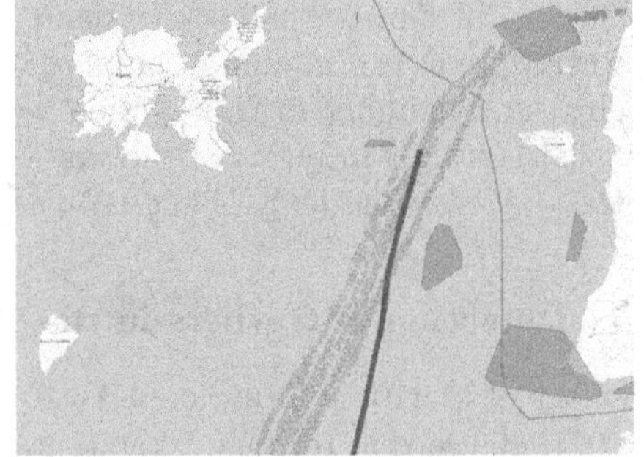

(d) A trajectory that slowly deviates from its course.

Fig. 5. Outliers detected by the proposed trajectory clustering.

Figure 5a illustrates a vessel's trajectory towards Naples, Italy. During its voyage the vessel makes a small circle and then continues its journey as before. Since its heading and speed changed dramatically the points in the circle (i.e. white) are considered outliers. Figure 5b illustrates the maritime traffic from the west to east, near Sicily, Italy. The trajectories from multiple vessels are grouped in the same cluster, since they share the same course and speed values and are drawn with the same colour (i.e. magenta). The centroid of this cluster has a heading of 102.3° and a speed of 13.91 knots. However, the part of the trajectory of a vessel that deviates from the normal route, starts heading to the north and after a while follows the same direction as before is marked with blue and yellow dots, since it moved to a different cluster. The blue cluster centroid has a heading of 30.2° and a speed of 1.2 knots (with a standard deviation of 5.25), whereas the respective centroid for the yellow cluster has a heading of 11.4° and a speed of 1.2 knots (with a standard deviation of 2.75). The actual centroid values clearly indicate an outlying behavior from a vessel that changed its route in slow speed in an area where similar (i.e. cargo) vessels move in different speed and direction. In a different case, Fig. 5c visualizes the maritime traffic of cargo vessels in the Aegean sea, showing all the vessels heading to the port of Piraeus, passing south of the island of Evia and near the island of Andros, Greece. There are two distinct clusters in the plot: (i) a big one that contains the trajectory of vessels traveling from the north-east Aegean sea, with a centroid of 227.3° (stdev = 21.32) and 13.1 knots (stdev = 2.49) and (ii) one that contains vessels traveling from the north-west, with a centroid of 137.8° (stdev = 14.26) and 13.0 knots (stdev = 1.65). The two clusters eventually merge into one cluster when the vessels pass south of Evia. Almost hidden among the two clusters is a third smaller cluster (marked with purple points) which illustrates a large deviation of a vessel that does not follow the patterns of all vessels with the same destination. This last cluster has a centroid of 145.8° (stdev = 2) and 1.4 knots (stdev = 0.16). With the proposed clustering algorithm, this subtrajectory, which does not contain any large and sudden course change or a stop has been identified as an outlier. Finally, Fig. 5d shows the maritime traffic near the island of Lemnos, Greece. From the plot it is obvious that while all vessels follow a specific route (the same big cluster as in Fig. 5c), when they head towards the port of Piraeus, using similar speed and heading values, there is one vessel that slowly deviates (marked with blue colour) from the common route, for unknown reason. This outlier has an average heading of 192.7° and 9.8 knots speed. The comparison between the normal behavior (227.3°, with stdev = 21.32 and 13.1 knots, with stdev = 2.49) shows that this outlier moved much slower that all other cargo vessels too.

All the cases presented above, are extracted from a dataset of 1,716 cargo vessels, following a totally unsupervised method (clustering). As a consequence, it provided us with useful feedback on the applicability of the proposed method and on the type of deviations it can detect. However, the same methodology can be used as a basis for a supervised (classification) technique that will detect vessel deviations using pretrained cluster information.

5 Conclusion and Future Steps

In this work, we proposed a clustering technique, which can be used to enrich our previously proposed maritime traffic network [2] that can efficiently model the behavior of vessels using only free and openly transmitted AIS data. The modelling of the normal vessel behavior will allow us to further distinguish outliers in the trajectories that are of interest to the maritime authorities. In this work, we showcased a few real world examples which our model managed to accurately detect. Identifying specific cases of anomalous behavior [10,11,23,24] will allow us to fine-tune, improve and exploit the proposed unsupervised technique as a basis for a supervised model for the detection of events of interest in the maritime sector. As a future work, we intend to exploit the proposed network abstraction in order to identify events of interest to the maritime authorities. Besides the route deviation problem presented in the preliminary results, we are interested in identifying several other anomalies related to the maritime domain such as communication gaps, AIS spoofing and illegal activities, thus building a unified framework for anomaly detection in real-time. The evaluation of the future anomaly detection framework will take into account real-world incidents and will measure the detection latency in real-time.

Acknowledgements. This work has been developed in the frame of the MASTER and SmartShip projects, which have received funding from the European Union's Horizon 2020 research and innovation programme under the Marie Skłodowska-Curie grant agreement No. 777695 and No. 823916 respectively.

References

1. Montewka, J., Kujala, P., Ylitalo, J.: The quantitative assessment of marine traffic safety in the Gulf of Finland, on the basis of AIS data. Zeszyty Naukowe/Akademia Morska w Szczecinie, pp. 105–115 (2009)
2. Varlamis, I., Tserpes, K., Etemad, M., Júnior, A.S., Matwin, S.: A network abstraction of multi-vessel trajectory data for detecting anomalies. In: Proceedings of the Workshops of the EDBT/ICDT 2019 Joint Conference, Portugal, Lisbon, March 2019
3. Yap, P.: Grid-based path-finding. In: Cohen, R., Spencer, B. (eds.) AI 2002. LNCS (LNAI), vol. 2338, pp. 44–55. Springer, Heidelberg (2002). https://doi.org/10.1007/3-540-47922-8_4
4. (PDF) mR-V: Line Simplification through Mnemonic Rasterization
5. Fernandez Arguedas, V., Pallotta, G., Vespe, M.: Maritime traffic networks: from historical positioning data to unsupervised maritime traffic monitoring. IEEE Trans. Intell. Transp. Syst. **19**(3), 722–732 (2018)
6. Coscia, P., Braca, P., Millefiori, L.M., Palmieri, F.A.N., Willett, P.: Multiple ornstein-uhlenbeck processes for maritime traffic graph representation. IEEE Trans. Aerosp. Electron. Syst. **54**(5), 2158–2170 (2018)
7. Douglas, D.H., Peucker, T.K.: Algorithms for the reduction of the number of points required to represent a digitized line or its caricature. Cartogr. Int. J. Geogr. Inf. Geovis. **10**(2), 112–122 (1973)

8. Holst, A., Ekman, J.: Anomaly detection in vessel motion (2003)
9. Holst, A., et al.: A joint statistical and symbolic anomaly detection system: increasing performance in maritime surveillance. In: 2012 15th International Conference on Information Fusion, pp. 1919–1926, July 2012
10. Varlamis, I., Tserpes, K., Sardianos, C.: Detecting search and rescue missions from AIS data. In: 2018 IEEE 34th International Conference on Data Engineering Workshops (ICDEW), pp. 60–65, April 2018
11. Chatzikokolakis, K., Zissis, D., Spiliopoulos, G., Tserpes, K.: Mining vessel trajectory data for patterns of search and rescue. In: Proceedings of the Workshops of the EDBT/ICDT 2018 Joint Conference (EDBT/ICDT 2018), Vienna, Austria, 26 March 2018, pp. 117–124 (2018)
12. Laxhammar, R.: Anomaly detection for sea surveillance. In: 2008 11th International Conference on Information Fusion, pp. 1–8, June 2008
13. Laxhammar, R., Falkman, G., Sviestins, E.: Anomaly detection in sea traffic - a comparison of the Gaussian mixture model and the kernel density estimator. In: 2009 12th International Conference on Information Fusion, pp. 756–763, July 2009
14. Fu, Z., Hu, W., Tan, T.: Similarity based vehicle trajectory clustering and anomaly detection. In: IEEE International Conference on Image Processing 2005, vol. 2, pp. II–602. IEEE (2005)
15. Hexeberg, S., Flåten, A.L., Brekke, E.F., et al.: AIS-based vessel trajectory prediction. In: 2017 20th International Conference on Information Fusion (Fusion), pp. 1–8. IEEE (2017)
16. Etemad, M., Soares Júnior, A., Matwin, S.: Predicting transportation modes of GPS trajectories using feature engineering and noise removal. In: Bagheri, E., Cheung, J.C.K. (eds.) Canadian AI 2018. LNCS (LNAI), vol. 10832, pp. 259–264. Springer, Cham (2018). https://doi.org/10.1007/978-3-319-89656-4_24
17. Le Guillarme, N., Lerouvreur, X.: Unsupervised extraction of knowledge from S-AIS data for maritime situational awareness. In: Proceedings of the 16th International Conference on Information Fusion, pp. 2025–2032. IEEE (2013)
18. Pallotta, G., Vespe, M., Bryan, K.: Vessel pattern knowledge discovery from AIS data: a framework for anomaly detection and route prediction. Entropy **15**(6), 2218–2245 (2013)
19. Lee, J.-G., Han, J., Whang, K.-Y.: Trajectory clustering: a partition-and-group framework. In: Proceedings of the 2007 ACM SIGMOD International Conference on Management of Data, pp. 593–604. ACM (2007)
20. Etemad, M., Júnior, A.S., Hoseyni, A., Rose, J., Matwin, S.: A trajectory segmentation algorithm based on interpolation-based change detection strategies. In: Proceedings of the Workshops of the EDBT/ICDT 2019 Joint Conference, Portugal, Lisbon, March 2019
21. Tampakis, P., Pelekis, N., Andrienko, N., Andrienko, G., Fuchs, G., Theodoridis, Y.: Time-aware sub-trajectory clustering in Hermes@ PostgreSQL. In: 2018 IEEE 34th International Conference on Data Engineering (ICDE), pp. 1581–1584. IEEE (2018)
22. Ester, M., Kriegel, H.-P., Sander, J., Xu, X.: A density-based algorithm for discovering clusters a density-based algorithm for discovering clusters in large spatial databases with noise. In: Proceedings of the Second International Conference on Knowledge Discovery and Data Mining, KDD 1996, pp. 226–231. AAAI Press (1996)

23. Kontopoulos, I., Spiliopoulos, G., Zissis, D., Chatzikokolakis, K., Artikis, A.: Countering real-time stream poisoning: an architecture for detecting vessel spoofing in streams of AIS data. In: IEEE joint conferences DASC/PiCom/DataCom/CyberSciTech 2018, pp. 981–986, August 2018
24. Patroumpas, K., Alevizos, E., Artikis, A., Vodas, M., Pelekis, N., Theodoridis, Y.: Online event recognition from moving vessel trajectories. Geoinformatica **21**(2), 389–427 (2017)

Nowcasting Unemployment Rates with Smartphone GPS Data

Daisuke Moriwaki(✉)

CyberAgent, Inc., 1-12-1 Dogenzaka Shibuya-Ku, Tokyo, Japan
moriwaki_daisuke@cyberagent.co.jp

Abstract. Unemployment rate is one of the most important macroeconomic indicators. Central governments and market participants heavily rely on the index to assess the economies. However, official statistics of unemployment rate are released infrequently with substantial delay. Prediction of official statistics of labor market will be helpful for these authorities as well as private companies and even workers. In this paper, we combine massive location data coming from smartphones and mixed data sampling (MIDAS) techniques to predict current unemployment rate in Japan. We found GPS data is very useful to predict the status of labor markets.

Keywords: GPS data · MIDAS · Mixed data sampling · Location data · Unemployment rate · Time series analysis · Macroeconomic policy · Nowcasting · Forecasting

1 Introduction

Unemployment rate is widely considered as one of the most important macroeconomic indices. Most of the central governments put a very high priority on employment since the unemployment causes many problems including poverty, crimes and social instability. The importance of unemployment statistics is not limited to public sector. Market participants assess the macroeconomic environment with unemployment rates. Ordinary employees also benefit from knowing unemployment rates since they need to decide whether or not quit the current job based on the labor market condition.

Besides its importance, the index has a serious problem. Official statistics authorities such as the U.S. Bureau of Labor Statistics (BLS) and Eurostat publish unemployment rates on monthly basis. We can not notice acute changes in the labor market from the statistics. Furthermore, the statistics is usually published around a month after the expiry of the month. The delay is caused by the time spent on the distribution and collection of survey questionnaires, and data processing. For prompt macroeconomic policy intervention and efficient market functioning, important indices such as unemployment rates should be reported as quick as possible. Catastrophic economic event such as financial crisis

K. Tserpes et al. (Eds.): MASTER 2019, LNAI 11889, pp. 21–33, 2020.
https://doi.org/10.1007/978-3-030-38081-6_3

during 2007–2009 must be noticed by the government and important decision makers in the private sector before it gets seriously worse.

The need for the correct knowledge of current economic status leads to a large literature of nowcasting [1,2] and introduced to the real world. The FRB of Atlanta has been continuously updating real-time estimates of GDP using monthly economic statistics in their "GDPNow" website [3,4].

While the economic statistics rarely uses highly frequent data (e.g. hourly, daily, or weekly), a massive volume of high frequency data have become available. GPS log data is one instance. Many location-based apps such as maps, entertainment, game, and fitness collect users' geo-location information if the users give permissions. These data are primarily utilized for improvement of user experience as well as advertising, recommendation and business intelligence. However, we see fast-growing literature on statistical analysis using collected GPS logs in a variety of areas including prediction of demographics and preference, detection of home, mode detection, and population analysis to name a few [5–9].

Recently, "alternative data", or non-traditional data has been embraced in the non-academic area. In the financial industry, more and more market participants start using alternative data including geo-location data to make investment decisions. [10] Investors such as hedge funds predict sales by location data. [11] Rigorous consideration is needed in the field.

In this paper, we introduce GPS data to nowcasting literature and develop a unique model predicting current unemployment rates with GPS log. Our evaluation proves that GPS data has substantial predictive power for number of the unemployed persons. In the following sections, we first briefly review literature in Sect. 2, then explain our data in Sect. 3. Section 4 gives the detail of our model and Sect. 5 evaluates it. Finally Sect. 6 concludes.

2 Related Works

To the best of our knowledge, this is the first attempt to forecast unemployment rates with GPS data. Nowcasting of labor market statistics with alternative data has been actively studied since Varian and Choi [12] suggested the potential predictive power of search query data. The earliest attempts to forecast unemployment rate with search query reveal the predictive power of query data for labor market [13–15] and many studies follow (e.g. [16–18]). While most of the papers utilize ARIMA-type models, Onorante and Koop [19] apply Dynamic Model Selection/Averaging and Scott and Varian [20] develop the Bayesian structural time series model.

The present work considers mixed data sampling (MIDAS) scenario pioneered by Ghysels et al. [21] in which the high frequency data is used to forecast infrequent data.The idea of MIDAS is to represent frequent data in a parsimonious way. A natural extension is a situation where high dimensional (large p) and high frequency predictive variables are present in small sample (smaller N). Various models combine feature selection techniques and MIDAS are proposed [22–24]. Recently Uematsu and Tanaka [25] showed a simple penalized regression without

MIDAS technique performs well for GDP forecasting with high frequent data. While these research focus on monthly official statistics as high frequent data and quarterly data (usually GDP) as target. The present paper extends MIDAS to much more high frequent alternative data.

Moreover, unlike existing models, our model is unique in its purely static form, which reveals the predictive power of GPS itself.

3 Data

In this section, we explain the data for the target (unemployment rates) and the predictor (GPS logs) in detail.

3.1 The Unemployment Rate

The unemployment rate is defined as "the number of unemployed persons as a percentage of the total number of persons in the labour force" [26]. In mathematical form,

$$u = y/l, \tag{1}$$

where y and l denote the number of unemployed persons and labor force.

The number of unemployed persons and persons in the labor force are usually surveyed by the government on monthly basis. In Japan, monthly *Labor Force Survey* takes the role. The survey collects information about labor status of approximately 40,000 households during the last week of each month. To estimate the number of unemployed persons, we take advantage of the fact that they have strong incentives to go to public employment service offices. It is mandatory for Japanese unemployed workers to visit one of public employment services offices to become eligible for unemployment insurance benefits. Furthermore, they have to visit the office at least once a month to maintain their eligibility [27]. We can easily presume more visitors implies more unemployed persons.

Once we get the number of unemployed persons, we need the number of labor force to divide it. Unfortunately finding clues for the number of labor force from the GPS data is not very easy. However, labor force is far less volatile and thus the prediction error is relatively small. A simple ARIMA model produces accurate predictions with the RMSE of 0.22 million and MAE of 0.18 million when the mean of labor force is 66 million.

In short, we estimate seasonally-adjusted unemployment rate u_t^{SA} as,

$$\hat{u}^{SA} = \hat{y}^{SA}/\hat{l}^{SA}, \tag{2}$$

$$\hat{y}^{SA} = \hat{y}^{\text{GPS}}/s^U \tag{3}$$

where s^U is seasonality index for unemployed persons. In the following sections, we first focus on the estimation of y rather than u. Resulting estimates of unemployment rate is shown in Sect. 5.

3.2 The GPS Data

Throughout this paper, we heavily rely on GPS logs from smartphones. Many mobile apps collect users' geographical location information to improve their services when the users give permission. We use completely annonymized version of GPS data taken from Jan 2016 to April 2019 (40 months). The data consists of four columns: hashed id, latitude, longitude and timestamp. We count the number of app users who possibly visit each employment service office daily basis. The resulting data consists of N (the number of offices) $\times$ D (the number of days) data points. We decide a person visits an employment service office when one or more logs are found within specific areas covering each office (Fig. 1). Since mobile phone determines its location based on the signals from GPS satellites, the accuracy deteriorates during a user is inside buildings or surrounded tall buildings due to the reflections of signals (multi-path). Furthermore, the logs are recorded infrequently to reduce battery consumption. To circumvent risk that we fail to count the person inside building due to the inaccuracy and infrequency of the nature of GPS data, the areas need to have some buffer outside the building.

The areas are set based on the size of the offices. To get size of the buildings we applied OpenCV [29] to map. The number of logs represent the number of visitors who has installed and given permission to specific app(s). The numbers are affected by whether the smartphone is turned on/off, whether the apps are turned on or not, and whether GPS logs are accurate. Moreover, note that visitors are not always unemployed persons. Visitors include consultants of the office, HR staffs from companies and other related people. Nevertheless, the numbers are expected to include some information about the number of unemployed. The counts are normalized by dividing by the total number of the daily unique users to mitigate the effect from the change in data volume.

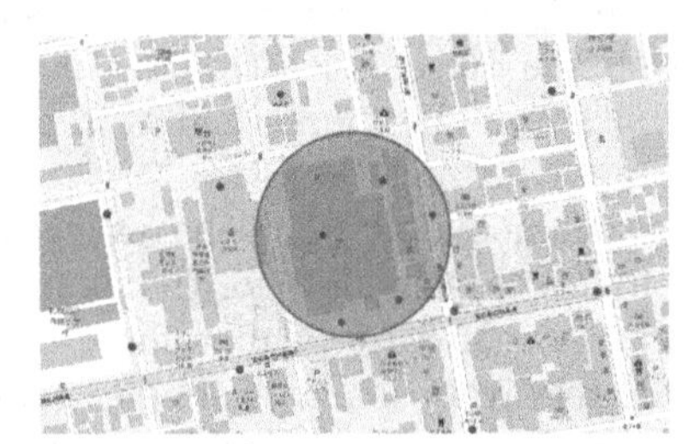

Fig. 1. Image of GPS data and an employment service office. Logs (red points) found in the green-colored area are counted. (Color figure online)

4 Nowcasting Model

In this section, we set up a nowcasting model and explain the estimation. Algorithm 1 summarizes the whole procedure.

Algorithm 1. Nowcasting of The Number of Unemployed Persons

Require: Daily GPS data $\{x^H_{n,1-\frac{29}{30}}, x^H_{n,1-\frac{28}{30}}, \cdots, x^H_{n,\bar{t}-\frac{\bar{d}}{30}}\}_{n\in\{1,N\}}$

Monthly data for the number of unemployed persons from official statistics $\{y_t\}_{t=1}^{\bar{t}-\text{lag}}$, lag = {1,2}

Set $\{\phi_i\}_{i=0}^{30}$ according to normalized beta distribution $(k, \alpha, \beta) = (1, 1.3, 1)$, where k is normalization factor

for $n = 1, \cdots, N$ **do**

  Impute $\{x^H_{n,\bar{t}-\frac{\bar{d}-1}{30}}, x^H_{n,\bar{t}-\frac{\bar{d}-2}{30}}, \cdots, x^H_{n,\bar{t}}\}$ with univariate ARIMA model

  for $t = 1, \cdots, \bar{t}$ **do**

    calculate $x^L_{n,t} = \sum_{i=0}^{d-1} \phi_i x^H_{n,t-i/30}$.

  end for

  Discard $\{x^L_{n,t}\}_{t=1}^{\bar{t}-\text{lag}}$ if $corr(\{x^L_{n,t}\}_{t=1}^{\bar{t}-\text{lag}}, \{y_t\}_{t=1}^{\bar{t}-\text{lag}}) \leq 0.3$

end for

Learn f from $(y_t, \{x^L_{n,t}\}_n, \mathbf{z}_t)_{t\leq\bar{t}-\text{lag}}$

Forecast $\hat{y}_{\bar{t}} = f(x^L_{n,\bar{t}}, \mathbf{z}_{\bar{t}})$

4.1 The MIDAS Model

Since unemployment rates are monthly statistics, it is not straightforward to develop a predictive model using daily data. As discussed in Sect. 2, such a situation is called "mixed data sampling" or MIDAS in short. We employ a most simple variant of MIDAS models, "bridge equation". Ghysels and Marcellino (2018) [30] provides detailed explanation on MIDAS models. Notations used in this paper are based on the book. Suppose y_t is a monthly (low frequency) outcome variable to be predicted and $x^H_{n,t-i/d}$ is N daily (high frequency) feature variables. The two variables themselves are not compatible with each other. We need to "bridge" high frequency data $x^H_{n,t}$s to low frequency x^L_t. That is,

$$x^L_{n,t} = \sum_{i=0}^{30} \phi_i x^H_{n,t-i/30} \tag{4}$$

where ϕ_is are positive scalars holds $\sum_{i=0}^{30} \phi_i = 1$. Hereafter we assume every month has 31 days regardless of the month. We pad zeros to the first d^* days for months with fewer than 31 days. For example, non-Olympic year February (28 days) goes like $(0, 0, 0, \text{1st day}, \cdots, \text{28th day})$. Then with a suitable machine learning model f, one can forecast y_t.

$$\hat{y}_t = f(x^L_{1,t}, \cdots, x^L_{N,t}, \mathbf{z}_t), \tag{5}$$

where $\mathbf{z_t}$ includes month and year.

4.2 Estimation of Parameters

We have two sets of parameters to be estimated. One is a vector of ϕ which transform daily data to monthly data. The other is parameters in the model

f, which gives prediction of y from x. In MIDAS literature, weight vector ϕ is chosen from several options. [21] Here we tried linear scheme $\phi_i = 1/31$ and normalized beta $\phi_i = \frac{beta(i/30,\alpha,\beta)}{\sum_{i=0}^{30} beta(i/30,\alpha,\beta)}$ where $beta(x,\alpha,\beta) = \frac{x^{\alpha-1}(1-x)^{\beta-1}\Gamma(\alpha+\beta)}{\Gamma(\alpha)\Gamma\beta}$. We go with normalized beta as it outperforms linear scheme. β governs the peak of the weights and α governs the slope of the weights (see Fig. 2). Since official monthly labor survey collects data during the last week of the month, it is reasonable to set $\beta = 1$. Finally α is chosen according to the resulting RMSE and MAE by grid searching.[1]

For forecasting model f we need to consider that the number of employment service offices (544) is much larger than the number of data points (40 months). This means standard MIDAS regression is not applicable[2]. We pick up standard Random Forest and L1-regularized least squares (LASSO). More flexible regression models such as SVM and neural nets are not suitable for our short time series data. Furthermore, when evaluating the model, training data gets much more shorter. Our model should not learn data from the future. We evaluate the model on data from May 2018 to Apr 2019. It leaves only 28 months to learn when evaluated at May 2018. Random forest out-perform LASSO for the most of the cases, we go with random forest[3].

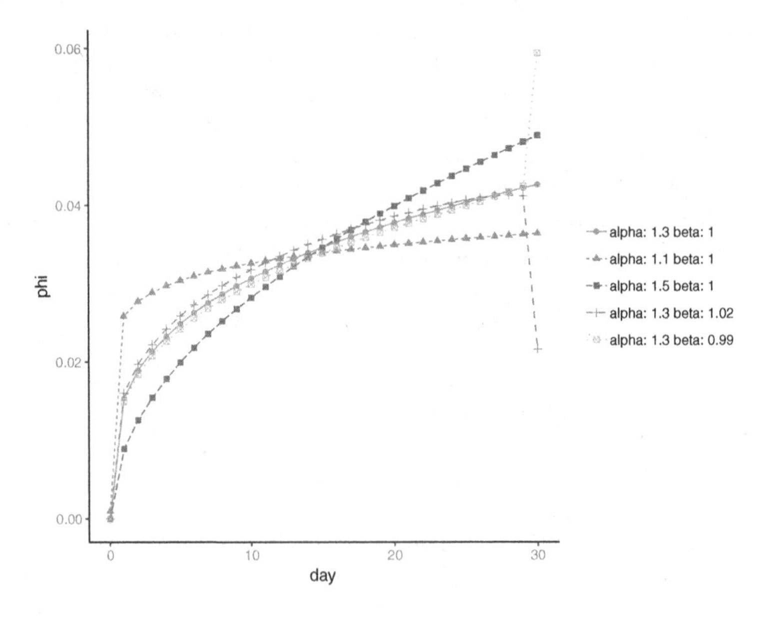

Fig. 2. Values of ϕ by parameters.

[1] The weights are generated by R package **midasr**.

[2] R package **midasr** does not have implementation for regularization.

[3] We used R package **ranger** [28].

4.3 Imputation of Missing Data

Since our goal is to nowcast unemployment rate as quick as possible, ideally we want to estimate unemployment rate any day in month. Suppose we are on June 28th 2019, the day official statistics for May is released (Fig. 3). One wants to forecast unemployment rate for June 2019. This is called "one month ahead prediction" since it predicts unemployment for one month ahead. Then we need impute missing GPS data for three days (28th, 29th and 30th). We use standard ARIMA models to impute missing GPS data. The models are run separately for each office. Parameters are automatically selected by **auto.arima** of R package **forecast**. At most five days imputation suffices for one month ahead prediction.

What if we want to conduct two month ahead prediction? If you are in the middle of the month (e.g. July 15th) then you need to impute 16 days of GPS logs. We check how the imputation affects prediction performance in the experiment.

Alternatively, we can estimate model without imputation by using only available data. However, the number of available days of data changes day by day and we need a lot of predictive models to be estimated (p. 462 in [30]). Here we resort to one predictive model with imputation for simplicity.

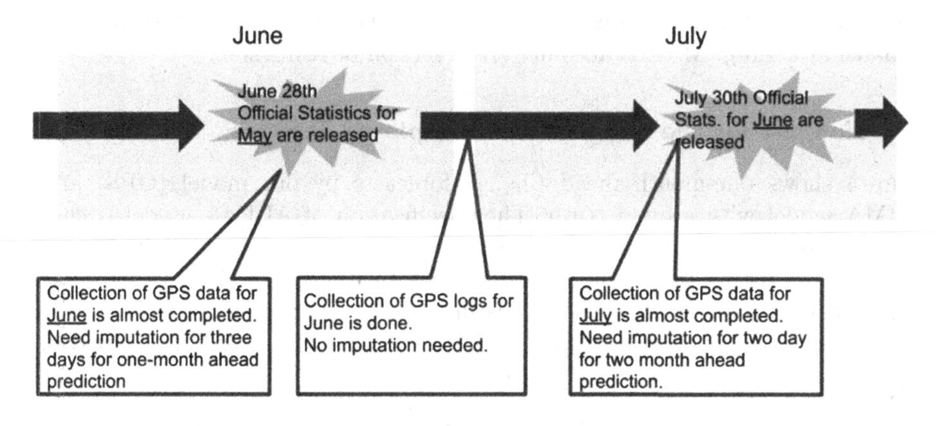

Fig. 3. Publication schedule and Imputation

4.4 Feature Selection

As already discussed in Sect. 2, feature selection is another important task here. Although random forest automatically selects informative feature variables, heuristic feature selection will benefit. Since the number of visitors to each office are expected to positively correlate with the number of unemployed persons, offices with negative correlation should be dominated by noise. We first calculate correlation of the data from each offices and official statistics for the number of unemployed persons in *training* period. Then we discard data from the offices with correlation smaller than 0.3. This procedure, however, leaves more than a hundred of offices.

5 Evaluation

In this section, we evaluate our predictive model by comparing baseline models. We first examine model for the number of unemployed persons and then one for unemployment rates. Throughout the section we utilize root of mean square error (RMSE) and mean absolute error (MAE) of the model for 12 months of rolling forecast. RMSE is defined as

$$\text{RMSE} = \left(\sum_{t=\text{May2018}}^{\text{Apr2019}} (y_t - \hat{E}[y_t|\hat{x}_t^L])^2 \right)^{1/2} \quad \text{for GPS} \tag{6}$$

$$\text{RMSE} = \left(\sum_{t=\text{May2018}}^{\text{Apr2019}} (y_t - \hat{E}[y_t|y_{t-lag}, \dots, y_1])^2 \right)^{1/2} \quad \text{for ARIMA} \tag{7}$$

where y_t denotes the ground truth taken from official statistics and lag indicate the number of steps of the forecasts. Note that $\hat{x}^L$ is estimated using information available at $t - lag$. MAE is absolute error version of RMSE.

5.1 Nowcast for the Number of Unemployed Persons

Figure 4 shows one-month-ahead ($\hat{y}_{t|t-1}$) forecasts by our model (GPS) and ARIMA model with ground truth. The specification of ARIMA model is chosen by **auto.arima**. It takes seasonality into account. As we've already seen in Sect. 3.2, forecasting before the end of the month needs some imputation. The right panel shows forecast without missing data while the left shows forecast with three days missing. Since there is a substantial delay in official statistics, imputation is not necessary in the most of the cases.

In general, the GPS model (green solid line) well predicts true values (blue dot-dashed) with several exceptions (May 2018, Jan 2019, and Mar 2019). One of the stinking features of the GPS model is its smoothness of the prediction. Compared to ARIMA (red dashed line), the predictions of the GPS model are far less volatile. In particular, ARIMA tends to mimic the level of the last month while the GPS model does not. This is reasonable because GPS model is a simple static model and does NOT have an autoregressive characteristic. The shape of the prediction by GPS model seems too smooth. It fails to predict some dips in the ground truth. However, economists want to see the trend of the economy rather than the short-term fluctuation. That's why economists prefer moving averaged indicators. GPS model accurately predict downward trend in unemployment.

Figure 5 shows the two months ahead forecasts ($y_{t|t-2}$). The right hand panel shows the forecasts based on data with five days missing while the left miss

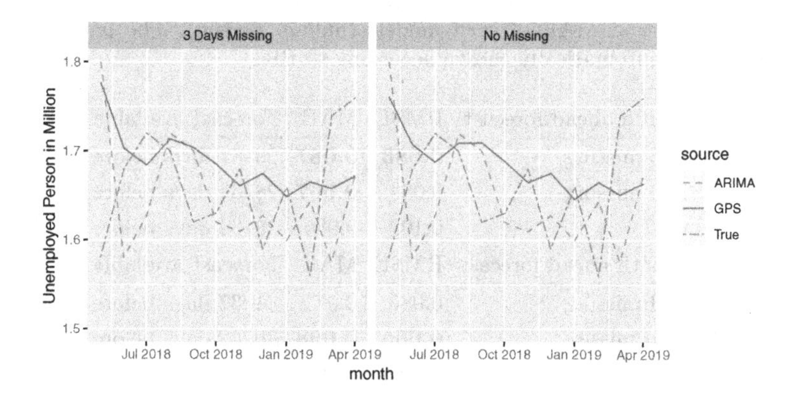

Fig. 4. One month ahead forecast ($y_{t|t-1}$) by proposed model (GPS), ARIMA model, and ground truth (true)

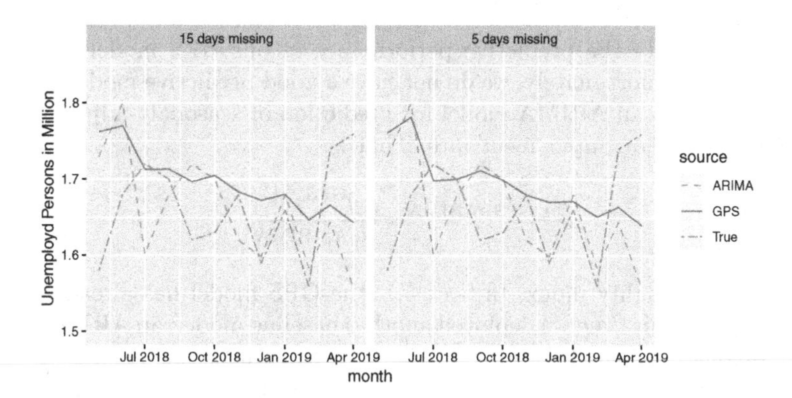

Fig. 5. Two month ahead forecast ($y_{t|t-2}$) by proposed model (GPS), ARIMA model, and ground truth (true)

fifteen days of GPS log. Compared with one month ahead forecasts, two months ahead forecasts shows larger errors for several months (Feb 2019 and Oct 2018). However, the results are much better than ARIMA model.

Table 1 summarizes the performance of the models. GPS models out-perform ARIMA model both one month ahead and two months ahead forecasts. Also the period of imputation seems not to affect the performance. Even model trained on data with 15 days imputation outperform ARIMA model. Also, the accuracy is almost same for one or two month ahead forecasts. The only difference between one month ahead and two month ahead GPS models is that two month ahead models do not learn data of just before the target month (i.e. y_{t-1}). Learning the last month of the target month might not be so important.

Table 1. RMMSE and MAE of each models (million person). The parameters of ARIMA model are automatically chosen according to BIC.

One month ahead forecast	RMSE	MAE	Forecast available
GPS, no missing	**0.083**	**0.067**	26–31 days before
GPS, 3d missing	0.084	**0.067**	28–34 days before
ARIMA	0.102	0.086	28–34 days before
Two month ahead forecast	RMSE	MAE	Forecast available
GPS, 5d missing	0.086	0.07	31–37 days before
GPS 10d missing	**0.085**	**0.068**	36–42 days before
GPS 15d missing	**0.085**	0.069	41–47 days before
ARIMA	0.101	0.083	56–64 days before

5.2 Forecasts for Unemployment Rates

Finally, we evaluate the predictive performance of our GPS model for unemployment rates. Unfortunately, we do not have a good predictive model for labor force. We resort to an ARIMA model for prediction of seasonaly-adjusted labor force and estimate unemployment rate. That is,

$$\hat{u}^{\mathrm{SA,GPS-ARIMA}} = \frac{\hat{y}^{\mathrm{GPS}}/s^U}{\hat{l}^{\mathrm{SA,ARIMA}}}, \tag{8}$$

where s^U is seasonality index. In Sect. 5.1, the GPS model has already beaten ARIMA model. This time we deployed another baseline model: an ARIMA model directly predicts seasonally adjusted unemployment rates. The results (Table 2) show our GPS-ARIMA model is inferior to the ARIMA model for one month prediction horizon ($\hat{u}_{t|t-1}$) but is competitive for two month prediction horizon ($\hat{u}_{t|t-2}$). As shown in Fig. 6, the up-and-down of the ground truth is better

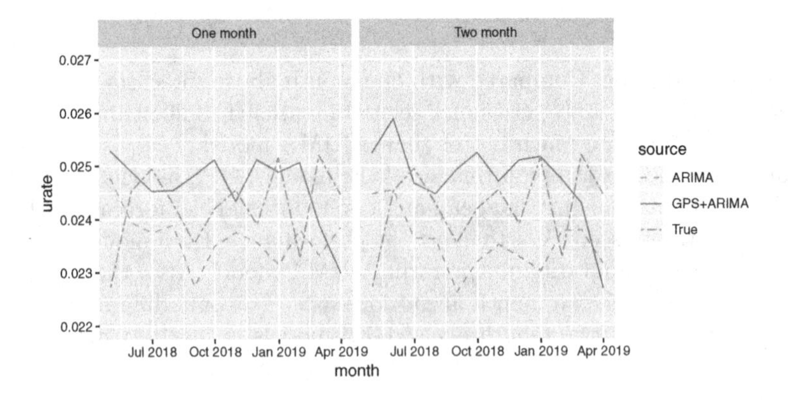

Fig. 6. Forecasting of unemployment rates

Table 2. Performance of predictive model for unemployment rates. RMSE/MAEs are inflated by 1,000. For example, 1.0 of MAE implies 0.1% mean absolute error.

One month ahead	RMSE × 1000	MAE × 1000
GPS, no missing	1.22	1.00
ARIMA	**1.16**	**0.99**
Two month ahead	RMSE × 1000	MAE × 1000
GPS 10 days missing	1.25	1.02
GPS 15 days missing	1.22	**0.97**
ARIMA	**1.19**	1.05

predicted by ARIMA while the absolute values are better predicted by GPS-ARIMA (e.g. Jul 2018, Nov 2018, Jan 2019).

The disappointing result is actually no surprise. The existing literature shows that the predictive power of alternative data is sometimes weak. [16,19] Also, the better predictive model for labor force could improve the results.

6 Conclusion

In this paper, we examined the usefulness of GPS log data for nowcasting for unemployment rates. First we prove that model using GPS data without the lagged dependent variable out-performs a standard ARIMA model for prediction of the number of unemployed persons. Then we found that the a combination of GPS and ARIMA model is only competitive for longer prediction horizon when applied to unemployment rates. The predictive performance could be improved by several ways. First, as described in Sect. 2, various modern techniques for MIDAS and high dimensional data are available. Second, using GPS data as an independent variable in an autoregressive model is another good candidate. Third, more sophisticated treatment for GPS log is expected to improve the quality of the data. Counting log is simple but the literature on GPS trajectories suggests many other technique to improve accuracy. Nevertheless, we hope the paper presents new idea for both nowcasting of economic statistics and utilization of GPS data.

References

1. Giannone, D., Reichlin, L., Small, D.: Nowcasting: the real-time informational content of macroeconomic data. J. Monet. Econ. **55**, 665–676 (2008). https://doi.org/10.1016/j.jmoneco.2008.05.010
2. Bańbura, M., Giannone, D., Modugno, M., Reichlin, L.: Now-casting and the real-time data flow. In: Handbook of Economic Forecasting, pp. 195–237. Elsevier (2013). https://doi.org/10.1016/B978-0-444-53683-9.00004-9
3. Federal Reserve Bank of Atlanta: GDPNow. https://www.frbatlanta.org/cqer/research/gdpnow.aspx. Accessed 11 June 2019

4. Higgins, P.C.: GDPNow: a model for GDP "Nowcasting". SSRN Electron. J. (2014). https://doi.org/10.2139/ssrn.2580350
5. Sangaralingam, K., Verma, N., Ravi, A., Datta, A., Chugh, V.: Predicting age & gender of mobile users at scale - a distributed machine learning approach. In: 2018 IEEE International Conference on Big Data (Big Data), pp. 1817–1826. IEEE, Seattle, WA, USA (2018). https://doi.org/10.1109/BigData.2018.8621942
6. Ravi, A., Sangaralingam, K., Datta, A.: Predicting consumer level brand preferences using persistent mobility patterns. In: 2018 IEEE International Conference on Big Data (Big Data), pp. 1986–1991. IEEE, Seattle, WA, USA (2018). https://doi.org/10.1109/BigData.2018.8622225
7. Vanhoof, M., Reis, F., Ploetz, T., Smoreda, Z.: Assessing the quality of home detection from mobile phone data for official statistics. J. Off. Stat. **34**, 935–960 (2018). https://doi.org/10.2478/jos-2018-0046
8. Siła-Nowicka, K., Vandrol, J., Oshan, T., Long, J.A., Demšar, U., Fotheringham, A.S.: Analysis of human mobility patterns from GPS trajectories and contextual information. Int. J. Geogr. Inf. Sci. **30**, 881–906 (2016). https://doi.org/10.1080/13658816.2015.1100731
9. Shimosaka, M., Hayakawa, Y., Tsubouch, K.: Spatiality preservable factored Poisson regression for large-scale fine-grained GPS-based population analysis. In: AAAI 2019, The Thirty-Third AAAI Conference on Artificial Intelligence (AAAI-19), January 2019
10. Opimas: Alternative Data - The New Flontier in Asset Management. http://www.opimas.com/research/217/detail/
11. Advan Research: Advan Location White Paper. https://www.advan.us/research.html
12. Choi, H., Varian, H.: Predicting the present with Google trends. Econ. Rec. **88**, 2–9 (2012). https://doi.org/10.1111/j.1475-4932.2012.00809.x
13. Askitas, N., Zimmermann, K.F.: Google econometrics and unemployment forecasting. Appl. Econ. Q. (formerly: Konjunkturpolitik) **55**, 107–120 (2009)
14. D'Amuri, F., Marcucci, J.: "Google It!" forecasting the US unemployment rate with a Google job search index. SSRN Electron. J. (2010). https://doi.org/10.2139/ssrn.1594132
15. Suhoy, T.: Query Indices and a 2008 Downturn: Israeli Data, vol. 34 (2009)
16. Pavlicek, J., Kristoufek, L.: Nowcasting unemployment rates with Google searches: evidence from the visegrad group countries. PLoS ONE **10**, e0127084 (2015). https://doi.org/10.1371/journal.pone.0127084
17. Anvik, C., Gjelstad, K.: Just Google it. Forecasting Norwegian unemployment figures with web queries (2010)
18. Naccarato, A., Falorsi, S., Loriga, S., Pierini, A.: Combining official and Google trends data to forecast the Italian youth unemployment rate. Technol. Forecast. Soc. Chang. **130**, 114–122 (2018)
19. Onorante, L., Koop, G.: Macroeconomic nowcasting using Google probabilities. In: Proceedings of the 1st International Conference on Advanced Research Methods and Analytics. Universitat Politècnica València (2016). https://doi.org/10.4995/CARMA2016.2016.4213
20. Scott, S.L., Varian, H.R.: Predicting the present with Bayesian structural time series. Int. J. Math. Model. Numer. Optim. **5**, 4–23 (2014)
21. Ghysels, E., Sinko, A., Valkanov, R.: MIDAS regressions: further results and new directions. Econ. Rev. **26**(1), 53–90 (2007)
22. Marsilli, C.: Variable selection in predictive MIDAS models. Banque de France Working Paper No. 520 (2014). https://doi.org/10.2139/ssrn.2531339

23. Siliverstovs, B.: Short-term forecasting with mixed-frequency data: a MIDASSO approach. Appl. Econ. **49**(13), 1326–1343 (2017)
24. Mogliani, M.: Bayesian MIDAS penalized regressions: estimation, selection, and prediction. arXiv:1903.08025 (2019)
25. Uematsu, Y., Tanaka, S.: High-dimensional macroeconomic forecasting and variable selection via penalized regression. Econ. J. **22**, 34–56 (2019). https://doi.org/10.1111/ectj.12117
26. International Labour Organization: Unemployment Rate. https://www.ilo.org/ilostat-files/Documents/description_UR_EN.pdf. Accessed 11 June 2019
27. Employment Security Bureau, Ministry of Health, Labour, and Welfare: Procedures of Employemnt Insurance in Japanese. https://www.hellowork.go.jp/insurance/insurance_procedure.html. Accessed June 11 2019
28. Wright, M.N., Ziegler, A.: ranger: a fast implementation of random forests for high dimensional data in C++ and R. J. Stat. Softw. **77** (2017). https://doi.org/10.18637/jss.v077.i01
29. Bradski, G.: The OpenCV library. Dr. Dobb's J. Softw. Tools **25**, 120–125 (2000)
30. Ghysels, E., Marcellino, M.: Applied Economic Forecasting Using Time Series Methods. Oxford University Press, Oxford (2018)

Online Long-Term Trajectory Prediction Based on Mined Route Patterns

Petros Petrou[1(✉)], Panagiotis Tampakis[1], Harris Georgiou[1], Nikos Pelekis[2], and Yannis Theodoridis[1]

[1] Department of Informatics,
80 Karaoli & Dimitriou str., P.O. 18534, Piraeus, Greece
{ppetrou,ptampak,hgeorgiou,ytheod}@unipi.gr
[2] Department of Statistics and Insurance Science, University of Piraeus,
80 Karaoli & Dimitriou str., P.O. 18534, Piraeus, Greece
npelekis@unipi.gr

Abstract. In this paper, we present a Big data framework for the prediction of streaming trajectory data by exploiting mined patterns of trajectories, allowing accurate long-term predictions with low latency. In particular, to meet this goal we follow a two-step methodology. First, we efficiently identify the hidden mobility patterns in an offline manner. Subsequently, the trajectory prediction algorithm exploits these patterns in order to prolong the temporal horizon of useful predictions. The experimental study is based on real-world aviation and maritime datasets.

Keywords: Trajectory prediction · Trajectory clustering · Mobility patterns · Big data

1 Introduction

Huge amounts of tracking data are being generated on a daily basis by GPS-enabled devices which are stored for analytics purposes. These constitute a rich source for inferring mobility patterns and characteristics, which, in turn, can be valuable to a wide spectrum of novel applications and services, from mobile social networking to aviation traffic monitoring. During the last years, such data have attracted the interest of data scientists, both in industry and academia, and are used to extract knowledge and useful features on what, how and for how long the moving entities are conducting individual activities related to specific circumstances. One of the most challenging tasks is to exploit these data by means of identifying historical mobility patterns, which, in turn, can gauge the procedure of discovering what the moving entities might do in the future. As a consequence, predictive analytics over mobility data have become increasingly important and are ubiquitous in many application fields [2,30,43].

The problem of predictive analytics over mobility data finds two broad categories of application scenarios. The first scenario involves cases where the moving entities are traced in real-time to produce analytics and compute short-term

K. Tserpes et al. (Eds.): MASTER 2019, LNAI 11889, pp. 34–49, 2020.
https://doi.org/10.1007/978-3-030-38081-6_4

predictions, which are time-critical and need immediate response. The prediction includes either location- or trajectory-related tasks. Short-term location and trajectory prediction facilitates the efficient planning, management, and control procedures while assessing traffic conditions in the road, sea and air transportation field. The latter can be extremely important in domains where safety, credibility and cost are critical and a decision should be made by considering adversarial to the environment conditions to act immediately. The second scenario involves cases where long-term predictions are important to identify cases which exceed regular mobility patterns, detect outliers and determine a position or a sequence of positions at a given time interval in the future. In this case, although response time is not a critical factor, it is still crucial in order to identify correlations between historical mobility patterns and patterns which are expected to appear. Long-term location and trajectory prediction can assist to achieve cost efficiency or, when contextual information is provided (e.g., weather conditions), it can ensure public safety in different transportation modes (land, sea, air).

As the maritime and the Air Traffic Management (ATM) domains have major impact to the global economy, a constant need is to advance the capability of systems to improve safety and effectiveness of critical operations involving a large number of moving entities in large geographical areas [22]. Towards this goal, the exploitation of heterogeneous data sources, which offer vast quantities of archival and high-rate streaming data, is crucial for increasing the computations accuracy when analysing and predicting future states of moving entities. However, operational systems in these domains for predicting trajectories are still limited mostly to a short-term look-ahead time frame, while facing increased uncertainty and lack of accuracy.

Motivated by these challenges, we present a Big data solution for online trajectory prediction by exploiting mined patterns of trajectories from historical data sources. Our approach offers predictions such as 'estimated flight of an aircraft over the next 10 min' or 'predicted route of a vessel in the next hour', based on their current movement and historical motion patterns in the area. The proposed framework incorporates several innovative modules, operating in streaming mode over surveillance data, to deliver accurate long-term predictions with low latency requirements. Incoming streams of moving objects' positions are cleansed, compressed, integrated and linked with archival and contextual data by means of link discovery methods.

This paper includes three main contributions: (a) we devise a big-data methodology/algorithm that solves the *Future Location Prediction* (FLP) problem in a effective and highly scalable way; (b) the design and implementation of our algorithm on top of state-of-the-art Big data technologies (namely Spark and Kafka); (c) extensive experimental study in large real datasets from the maritime and aviation domains. To the best of our knowledge, in contrast to related state-of-the-art systems [8,10] and research approaches [7], our approach is unique as a Big data framework capable of providing long-term trajectory predictions in an online fashion.

This paper is organized as follows. Section 2 presents the related work from the field of trajectory prediction and long-term future location prediction, especially from the maritime and aviation domains. Next, Sect. 3 describes the system overview and architecture of the proposed approach, as well as how this fits into the Big data scope. Section 4 presents the mobility pattern discovery module, in the form of a novel and scalable subtrajectory clustering Big Data solution, which is the first stage of this approach. Predictive models, which is the second stage, are described in Sect. 4.2. The experimental study in Sect. 5 includes datasets from both the maritime and the aviation domain. Finally, the conclusions and future aspects of this work are described briefly in Sect. 6.

2 Background

The trajectory of a moving object is defined as: $< (p_0, t_0), (p_1, t_1), ..., (p_i, t_i), ... >$, where p_i is the location of the object in d-dimensional space (typically, $d = 2$ or 3, for a movement in plane or volume, respectively) and t_i is the time this recording was made, with $t_i < t_{i+1}$ (i.e., the sequence is chronologically ordered).

Having this at hand, two main prediction-related problems can be stated for moving objects: Future Location Prediction (FLP) and Trajectory Prediction (TP) [14]. In these definitions we adopt the following terminology: symbols p and t refer to recorded or given locations and timestamps, respectively, whereas symbols p* and t* refer to (future) predicted locations and timestamps, respectively.

Problem Definition 1 **Future Location Prediction (FLP)**: Given (a) the incomplete trajectory $< (p_0, t_0), (p_1, t_1), ..., (p_{i-1}, t_{i-1}) >$ of a moving object o, consisting of its time-stamped locations recorded at past i time instances, and (b) an integer value $j \geq 1$, predict $< (p_i^*, t_i), ..., (p_{i+j-1}^*, t_{i+j-1}) >$, i.e., the objects's anticipated locations at the following j time instances.

Problem Definition 2 **Trajectory Prediction (TP)**: Given (a) the incomplete trajectory $< (p_0, t_0), (p_1, t_1), ..., (p_{i-1}, t_{i-1}) >$ of a moving object o consisting of its time-stamped locations recorded at past i time instances and (b) a target region R, predict $< (p_i^*, t_i), ..., (p^*, t^*) >$, where $p^* \in R$, i.e., the object's anticipated locations until it matches a point p^* in R (note: p^* may be never reached exactly).

Using these two baseline definitions for the FLP and TP tasks, a wide variety of algorithms can be employed to predict either sequences of future points (FLP) or the evolution of entire trajectories (TP). In the context of this work, the interest is focused specifically in TP or, complementary, to long-term FLP, i.e., with sufficiently large look-ahead time frames.

A typical example of a FLP method is presented in [38], where the authors propose TPR*-tree (index-based), which derives from TPR-tree, and exploits the characteristics of dynamic moving objects in order to retrieve only those which will meet specific spatial criteria within the given time interval, i.e., query window, in the future. Every moving object is represented by a Minimum Bounding Rectangle (MBR) along with a Velocity Bounding Rectangle (VBR). The proposed index integrates novel insertion and deletion algorithms to enhance performance and supports predictive spatio-temporal queries by specifying a query region q_r and a future time interval q_t and retrieving the set of objects that will intersect q_r at any timestamp $t \in q_t$.

The previous method can be considered as a FLP-based approach, mostly in the context of the long-term prediction. There is also a number of TP-based approaches that address the prediction task in a similar way. In theory, every FLP method can be transformed to a full TP model, given a specific granularity upon which the same method is applied iteratively. The main difference with 'pure' TP methods is that in this case the prediction errors are accumulated with each step (e.g. via multi-step Linear Regression) along the prediction track, thus making the predicted points increasingly error-prone. In contrast, TP methods forecast the complete trajectory as a whole, thus making each predicted point equally error-prone. Regarding en route climb TP, one of the major aspects of decision support tools for ATM, Coppenbarger [8] discusses the exploitation of real-time aircraft data, such as aircraft state, aircraft performance, pilot intent and atmospheric data for improving ground-based TP. The problem of climb TP is also discussed by Thipphavong, Schultz et al. [39], as it constitutes a very important challenge in ATM. In this work, an algorithm that dynamically adjusts modeled aircraft weights is developed, exploiting the observed track data to improve the accuracy of TP for climbing flights.

In the area of **stochastic approaches**, Ayhan and Samet [4] introduce a novel stochastic approach to aircraft trajectory prediction problem, which exploits aircraft trajectories, based on Hidden Markov Models (HMM), modeled in space and time by using a set of spatio-temporal data 4-D cubes (latitude, longitude, altitude, time) enriched by weather parameters. Gong and McNally [16] proposed a methodology for automated trajectory prediction analysis, specifically for splitting the process in separated stages according to the flight phases. The purpose is to identify flights, as described by actual radar tracks, which show unpredictable modifications of their aircraft intent and can be considered outliers. In another work by Ayhan and Samet [5], the authors investigate the applicability of the HMM for TP on only one phase of a flight, specifically the climb after takeoff. Moreover, they address the problem of incorporating weather conditions in their model, as they represent a major factor of uncertainty in all TP applications.

Regression and clustering are also two main areas of interest when applying machine learning methods in TP. Neural Networks (NN) have been proposed in various works as the core regression model for the task of TP. Le Fablec and Alliot [12] have introduced NNs for the specific problem of predicting an aircraft trajectory in the vertical plane, i.e., its altitude profile with the time. Cheng, Taoya, et.al. [6] employ a data mining statistical approach on the radar tracks of aircrafts to infer the future air traffic flows using Neural Networks (NN) and exploiting data grouped in seven 'weekday' categories for predicting the Estimated Time of Arrival (ETA) at designated fixes and airports as output. Leege, Paassen and Mulder [22] also address the specific TP task of predicting arrival routes and times via Generalized Linear Models (GLM), merging together air traffic following fixed arrival routes, meteorological data and two aircraft types.

In a very recent work of TP in aviation, Georgiou et al. [15] introduce flight plans, localized weather and aircraft properties as trajectory annotations that enable modelling in a space higher than the typical 4-D spatio-temporal. A multi-stage hybrid approach is employed for a new variation of the core TP task, the so called *Future Semantic Trajectory Prediction* (FSTP), including clustering the enriched trajectory data using a semantic-aware similarity function as distance metric. Subsequently, a separate predictive model is trained for each cluster, using a non-uniform graph-based grid that is formed by the waypoints of each flight plan. In practice, flight plans constitute a constrained-based training of each predictive model, one for each waypoint, independently. Various types of predictive models are tested, including HMM, linear regressors, regression trees and feed-forward NNs. The results show very narrow confidence intervals for the per-waypoint TP errors in HMM, while the more efficient linear and non-linear regressors exhibit 3-D spatial accuracy much lower than the current state-of-the-art, up to a factor of five compared to 'blind' TP for complete flights, in the order of 2–3 km compared to the actual flight routes.

Concerning mobility pattern discovery, the aim is to identify several types of collective behavior patterns among moving objects like the so-called flock pattern [20,41] and the notion of moving clusters [19]. A number of research efforts that emerged from the above ideas are the approaches of convoys [18,28], platoons [23], swarms [24], gathering pattern [42] and traveling companion [37]. Trasarti et al. [40] introduced "individual mobility patterns" in order to extract the most representative trips of a specific moving object, so that they can predict object's future locations. However, all of the aforementioned approaches are centralized and cannot scale to massive datasets. Towards this, the problem of convoy discovery in a distributed environment by employing the MapReduce programming model was studied both in [27]. An approach that defines a new generalized mobility pattern which models various co-movement patterns in a unified way and is deployed on a modern distributed platform (i.e., Apache Spark) to tackle the scalability issue is presented in [13].

Another line of research, tries to discover groups of either entire or portions of trajectories considering their routes. A typical strategy is to transform trajectories to a multi-dimensional space and then apply well-known clustering algorithms such as OPTICS [3] and DBSCAN [11]. Another approach is to define an appropriate similarity function and embed it to an extensible clustering algorithm [26]. Nevertheless, trajectory clustering is an "expensive" operation and centralized solutions cannot scale to massive datasets. Furthermore, [34] proposes a MapReduce approach that aims to identify frequent movement patterns from the trajectories of moving objects. In [17] the authors tackle the problem of parallel trajectory clustering by utilizing the MapReduce programming model and Hadoop. They adopt an iterative approach similar to k-Means in order to identify a user-defined number of clusters, which leads to a large number of MapReduce jobs.

However, discovering clusters of complete trajectories can overlook significant patterns that might exist only for portions of their lifespan. To deal with this, the authors of [21] propose TraClus, a partition-and-group framework for clustering 2-D moving objects which segments the trajectories based on their geometric features, and then clusters them by ignoring the temporal dimension. A more recent approach to the problem of subtrajectory clustering, is S^2T-Clustering [32], where the authors take into account the temporal dimension, and the segmentation of a trajectory takes place whenever the density of its spatiotemporal 'neighborhood' changes significantly. The segmentation phase is followed by a sampling phase, where the most representative subtrajectories are selected and finally the clusters are built "around" these representatives. A similar approach is adopted in [1], where the authors aim at identifying common portions between trajectories, with respect to some constraints and/or objectives, by taking into account the "neighborhood" of each trajectory. These common subtrajectories are then clustered and each cluster is represented by a pathlet, which is a point sequence that is not necessarily a subsequence of an actual trajectory. A different approach is presented in QuT-Clustering [31] and [35], where the goal is, given a temporal period of interest W, to efficiently retrieve already clustered subtrajectories, that temporally intersect W. To achieve this, a hierarchical structure, called ReTraTree (Representative Trajectory Tree) that effectively indexes a dataset for subtrajectory clustering purposes, is built and utilized.

The approach presented in this paper combines several aspects and ideas from the methods cited above, in order to develop a highly adaptive, long-term, Big data framework for FLP which is experimentally evaluated with datasets from both the maritime and the aviation domain. More specifically, this two-stage approach includes: (a) mobility pattern discovery from the historical movement of the moving objects; and (b) employ optimal estimations of FLP in the sense of maximum likelihood, as they are dictated by the identified patterns. Furthermore, some promising experimental results are presented for real datasets from both domains, as well as performance indicators for deployment in a Big data platform.

3 Overview of the Approach

In this section we describe the architecture of our proposed framework, which follows a typical lambda architecture [25] that combine streaming and batch layers to implement an end-to-end big data prediction solution. The proposed framework, as depicted in Fig. 2, consists of two main modules, namely, Pattern Extraction and Future Location Prediction (FLP). All modules are build on top of big data engines, so that they can be scalable and offer low latency. Kafka is used as an integration network for online toolboxes and a shared storage (i.e. Apache Hadoop HDFS) is used in order to update existing patterns or add new ones. Subsequently, the FLP module can "read" these patterns and execute the prediction pipeline.

At first, each moving object sends its location via traditional network protocols and then a Kafka producer collects all positions and pushes them to a Kafka topic. The Pattern Extraction module identifies "typical routes", in an offline manner. Finally, these "typical routes" are broadcast among all slaves and the FLP module combines them with the live incoming stream of data in order to predict the future location for each object (Fig. 1).

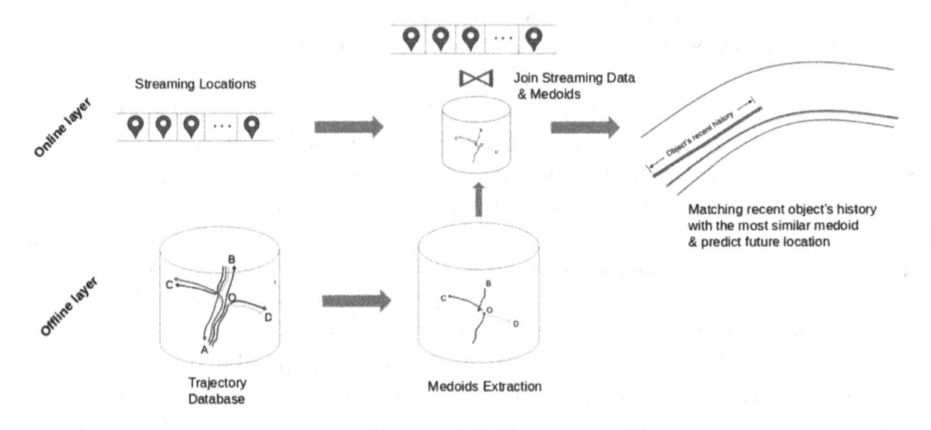

Fig. 1. Data workflow of the proposed framework

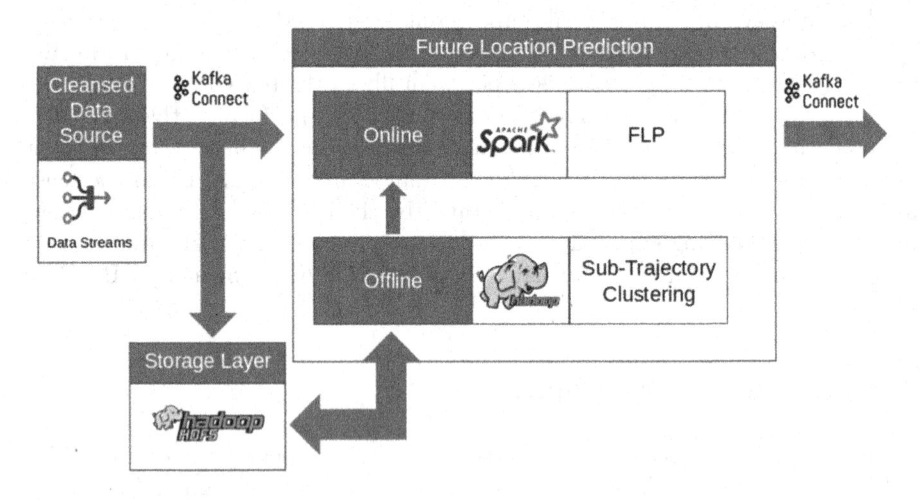

Fig. 2. Architecture of the proposed framework

4 Methodology

4.1 Offline Step: Mobility Pattern Extraction Based on Sub-trajectory Clustering

The goal of this module is to identify frequent patterns of movement that will assist the FLP module to increase the accuracy of the predictions. The research so far has focused mainly in methods that aim to identify specific collective behavior patterns among moving objects, such as flocks, convoys and swarms [44], or methods that try o identify patterns that are valid for the entire lifespan of the moving objects [9,26]. However, discovering clusters of entire trajectories can overlook significant patterns that might exist only for small portions of their lifespan. Furthermore, most of the approaches either operate at specific predefined temporal "snapshots" of the dataset and ignore the movement between these "snapshots" and/or ignore the temporal dimension and perform spatial-only clustering and/or assume that the length (number of samples) of the trajectories and the sampling rate is fixed, which is unrealistic. Another thing that should be taken into account when designing a prediction-oriented trajectory clustering algorithm, is that the resulting clusters should have a small extent in order for the predictions to be more accurate. Obviously, this, rules out a large number of approaches that perform density-based clustering which might lead to spatially extended clusters through expansion.

For the above reasons, the desired specifications that such a trajectory clustering algorithm should hold, in order to be able to predict the movement of future trajectories, are the following:

- Discovering of clusters of subtrajectories, instead of whole trajectories.
- Spatio-temporal clustering, instead of spatial only.
- Support of trajectories with variable sampling rate, length and with temporal displacement.
- Distance-based clustering.

There have been some approaches to deal with the problem of subtrajectory clustering in a centralized way [1,21,32], however, all the above subtrajectory clustering approaches are centralized and do not scale with the size of today's trajectory data, thus calling for parallel and distributed algorithms. For this reason, we utilize the work presented in [36], coined *DSC*, which introduces an efficient and highly scalable approach to deal with the problem of *Distributed Subtrajectory Clustering*, by means of MapReduce. More specifically, the authors of [36] split the original problem to three sub-problems, namely *Subtrajectory Join*, *Trajectory Segmentation* and *Clustering and Outlier Detection*, and deal with each one in a distributed fashion by utilizing the MapReduce programming model.

To elaborate more, the *Subtrajectory Join* step aims at retrieving for each trajectory $r \in D$, all the moving objects, with their respective portion of movement, that moved close enough in space and time with r, for at least some time duration. Subsequently, the *Trajectory Segmentation* step takes as input the result of the *Subtrajectory Join* step, which is actually a trajectory and its neighboring trajectories and targets at segmenting each trajectory $r \in D$ into a set of subtrajectories in a neighbourhood-aware fashion, meaning that a trajectory will be segmented whenever its neighbourhood changes significantly. Finally, the third step takes as input the output of the first two steps and the goal is to create clusters of similar subtrajectories and at the same time identify subtrajectories that are significantly dissimilar from the others (outliers).

For more details about the algorithms involved in *DSC* and an extensive experimental study, please refer to [36].

4.2 Online Step: On Long-Term Future Location Prediction

In this section, we describe how the FLP module takes advantage of an individual's typical movement (medoids from now on), based on the observation that moving objects often follow the same route patterns. This observation fits exactly in the maritime and aviation domain where vessels or airplanes have very strict routes between ports and airports, either implied due to route optimization (e.g. ship's fuel consumption) or explicitly required as official regulation (flight plans). The Future Location Prediction (FLP) module aims to make an accurate estimation of the next movement of a moving object within a specific look-ahead time frame.

Most approaches do not take advantage of any other historic data available, either from the object itself or other "similar" objects moving within the same area and context, making it susceptible to errors associated to noise, artifacts or outliers in the input. This results in inaccurate predictions and only with a short horizon (seconds or few minutes). A very different approach for the FLP problem is making the associated predictive models less adaptive but more reliable, by introducing specific "memory" based on historic data of an entire fleet of objects relevant to the context at hand. On the other hand, this requires a combination of historical and streaming data which is not a trivial task. A big challenge of our proposed framework is how to handle thousands of records efficiently in the context of online streaming data, join each object with the appropriate medoids and finally do all the necessary model calculations to produce predictions for the future locations of an object. In practice, several such medoids are pre-computed and stored in an efficient way (partitioned by object identifier), so that they can be retrieved on demand or even kept in-memory for several thousands of objects, making long-term FLP feasible in a large scale. This task is addressed by employing a Big data engine that is designed to conduct fast joins between streaming data and historical data. Spark module (SQL or Streaming) can efficient join historical and streaming data. Either with map-side-join (a.k.a broadcast join) or using Dataset (Spark structure) metadata to achieve extra optimizations. For example if the medoids can be sent to all workers (broadcast) at the initial phase, it is recommended to replicate medoids (create a local variable) in each worker and for each object in Map-Reduce phase we select its medoids to perform prediction. On the other hand, if the medoids' size cannot stored in each workers' memory, we partition the medoids by objects' identifier in order to have quick access for a specific object and create spark distributed structures that can be easily joined with Streaming data via Sparks SQL API.

Medoid matching: The first step tries to match the object's recent history with the medoids. More specifically, for all the medoids, we find the closest to the object's current trajectory. Algorithm 1 uses a spatiotemporal similarity function in order to find the best match. Prediction: The algorithm has already identified the last point from the best-matched, according to the previous stage. Then, it follows the medoid's points one by one until it reaches the prediction horizon.

The FLP-L approach described in brief above is inherently intuitive and self-explanatory. It relies on past routes of the same or similar objects in order to forecast how a specific object will move while it is already residing on a specific frequently-traversed route. The weighted similarity function between two spatiotemporal points $d(p,p') = \sqrt{w_1 \cdot (x-x')^2 + w_1 \cdot (y-y')^2 + w_2 \cdot (t-t')^2}$, was proposed in [29] and in our algorithm weights ratio is estimated by mean speed.

Algorithm 1 describes the prediction step in a more technical ςαυ. Actually, these steps is the Spark's map function after collecting streaming data in a certain (user-defined time window).

Algorithm 1: FLP-L Algorithm

```
Input  : current state (object's recent history), object's network, horizon,
         distance_threshold
Output: prediction path
min_dist ← Double.MaxValue;
best_match ← null;
foreach trajectory ∈ medoids do
    traj_medoid_distance ← SpatioTemporal Distance(current_state, trajectory);
    if traj_medoid_distance <min_dist AND traj_medoid_distance
    <distance_threshold then
        best_match ← trajectory;
        min_dist ← traj_medoid_distance;
    end
end
if best_match is not null then
    while prediction_path.getLast.getTimestamp <horizon AND
    best_match.hasNext do
        prediction_path.add(best_match.next());
    end
end
return prediction_path;
```

The above algorithm could be implemented in Spark Map-Reduce API as follows:

1. Receiving and parsing messages from input Kafka topic (map)
2. Reduce by object identifier over a window period
3. Join objects streaming data with the proper medoids.
4. Map partition (process each object for the current window) in order to perform prediction.

Step 3 is required only for the Dataset Join, otherwise (broadcast join) step 3 is performed inside step 4. Figure 3 illustrates an example of the FLP-L approach over a flight between Madrid and Barcelona, where the red points are the actual data and the blue points are the predictions.

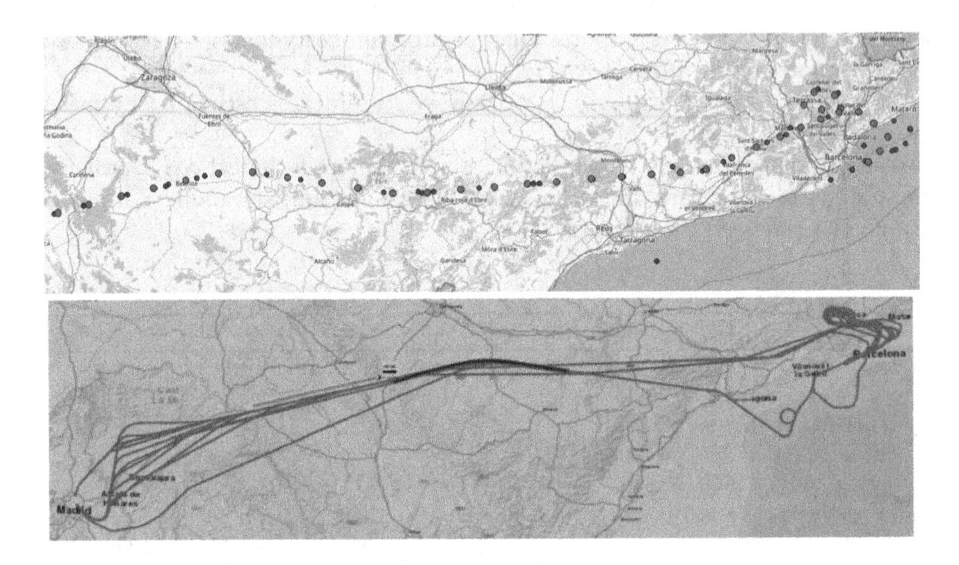

Fig. 3. Madrid - Barcelona flight example of the FLP-L approach. In the frst figure red points are real data and blue points are the predictions. In the second image red lines are medoids, gray line is the current window of a flight and the blue line is the predicted path (Color figure online)

5 Experimental Evaluation

5.1 Experimental Setup

In this section, we present the results of our experimental study. Our cluster consists of 10 nodes (1 master, 9 workers) with 5 executor cores per worker and 4 GB memory per worker. Input streams are provided by a Kafka topic and FLP-L is implemented on top of Spark SQL Streaming engine and Apache Yarn used as a resource manager. Spark SQL streaming tasks are processed using a micro-batch processing engine, which processes data streams as a series of small batch jobs thereby achieving low latency and exactly-once guarantees. Spark-Kafka integration is provided by Spark, but Spark tuning depends on parallelism, namely data partitioning and park Streaming integration for Kafka in our architecture provides simple parallelism and 1:1 correspondence between Kafka partitions and Spark partitions. His means that if we want the higher performance, we have to configure Spark to create the same partitions as Kafka and Kafka to have as many partitions as possible. For example, if input Kafka topic has 60 partitions, then the cluster must have at least 60 cores for the query to make progress and achieve the best performance. In our experiments we used one Kafka topic for each domain (aviation, maritime) with 60 partitions.

We conducted experiments against real datasets (IFS messages and AIS messages [33]). Table 1 summarizes some basic statistics about the input dataset.

Table 1. Dataset description

	Aviation	Maritime
Number of point	455000	16000000
Number of objects	680 flights	5055 MMSI
Spatial coverage	Spain (Madrid - Barcelona flights)	Brest area
Time span	April 2016 (one week)	6 months

5.2 Results

Based on the optimal Spark/Kafka configuration described in Fig. 4, the total delay originates almost entirely from the processing time, which asymptotically stabilizes at around 5 s. This essentially translates to 60,000 Kafka messages (points) per 10 s or 6,000 points/second, which corresponds to 8-min look-ahead window. In other words, with an average sampling rate of 5 s for each moving object, this system configuration of the FLP module can accommodate up to 30,000 moving objects with 5-s update and 8-min look-ahead predictions. It is also important to notice that scheduling time in Fig. 4, which is related with Spark-Kafka integration. Scheduling time with three workers overcome processing time because there are not enough resources (cores) in the Spark cluster in order to process input messages and Kafka input partitions. On the other hand, with six workers and above scheduler has enough resources to assign the planned tasks. This behaviour occurs because there are enough resources (cores) for executing Spark Tasks. On the other hand, with three workers there are not enough resources for the input messages for scheduling and the algorithm breaks. As described above, in this option a FLP approach is employed for exploiting the cluster medoids as "guidelines" for providing online predictions, e.g. as the actual flight evolves in real time. The general clustering method in this case is the same as described in Sect. 4. We use up to 14 clusters in order to perform future location prediction. The FLP module, uses sliding windows of 2 min of past positions in order to optimally match the most recent segment of the current trajectory to one of the available medoids, using a custom spatio-temporal similarity function. Then, the best-matched medoid is used as the maximum-likelihood trajectory evolution and the predicted positions are taken along its path for a specific (user-defined) look-ahead step.

Figure 5 illustrates the histogram of the horizontal error, i.e., the distribution of errors, for all the trajectories in the Aviation (Madrid/Barcelona) and Maritime (Brest Area) dataset and with spatial-only comparison (point-wise Euclidean distance). Specifically, they illustrate the boxplots of the per-complete-trajectory mean error for multiple look-ahead steps (1, 2, 4, 8, 16, 32 min). Additionally, the notation of the boxplot provides hints of the underlying error distributions, i.e., means, medians, upper/lower quartiles, non-outlier ranges, etc. These verify that the prediction errors are indeed in accordance with the expected shape of the distribution, i.e., a typical Extreme Value (EV) with medium/low skewness (Gaussian-like) towards the lower limit and an asymptotically decreasing right tail, i.e., accumulate and expand exponentially as the look-ahead span doubles.

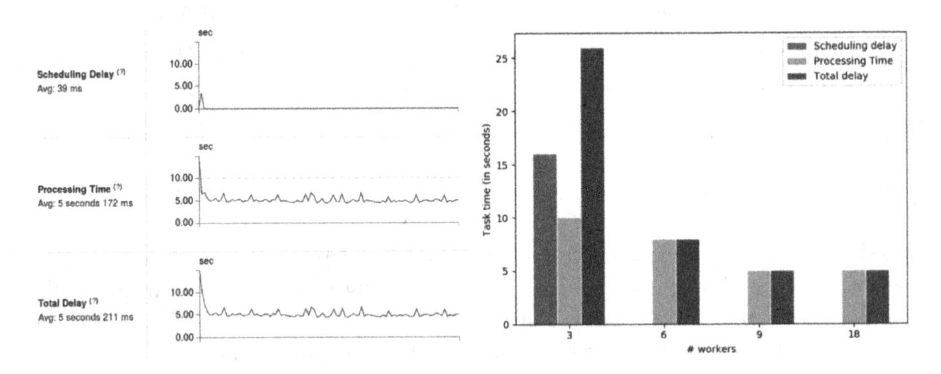

Fig. 4. Performance metrics for $16 \cdot 10^6$ points, $6 \cdot 10^3$ points/second, batch interval 10 s, 9 workers and 60 partitions.

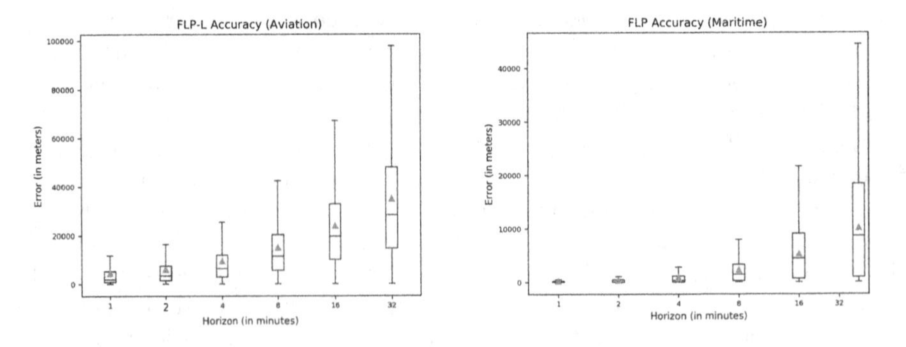

Fig. 5. Mean error for multiple look-ahead steps (1, 2, 4, 8, 16, 32 min), with custom spatio-temporal similarity function and with 90%-threshold outliers removed.

6 Conclusion

In this work, a novel approach was introduced for the long-term FLP problem (FLP-L). Our approach is based on purely data-driven extraction of mobility patterns, i.e. subtrajectory cluster medoids. This approach is generic enough to be applicable to various domain, such as in the aviation and maritime domain. It is important to emphasize that the proposed framework relies end-to-end in big data technologies The experimental results included here are focused primarily on the maritime domain, since the aviation is considered a more 'constrained' problem due to the fact that all flights are legally bounded to file and closely follow specific flight plans, i.e., the 'intended path' is much more specific and mandatory. Nevertheless, this framework is directly applicable and valid in the aviation domain too, especially since the medoids discovery is based upon some form of clustering to discover groups and common motion patterns, either with or without considering flight plans as input in the predictive models. The accuracy in both domains, as well as the performance results, prove that it is a very efficient and scalable Big data solution for real-world applications, easily adaptable to various other domains.

Acknowledgements. This work was partially supported by projects datACRON (grant agreement No 687591), Track&Know (grant agreement No 780754) and MASTER (Marie Sklowdoska-Curie agreement N. 777695), which have received funding from the EU Horizon 2020 R&I Programme. This research has also been co-financed by the European Regional Development Fund of the European Union and Greek national funds through the Operational Program Competitiveness, Entrepreneurship and Innovation, under the call RESEARCH-CREATE-INNOVATE (project code: T1EDK-3268).

References

1. Agarwal, P.K., Fox, K., Munagala, K., Nath, A., Pan, J., Taylor, E.: Subtrajectory clustering: models and algorithms. In: PODS, pp. 75–87 (2018)
2. Andrienko, G., Andrienko, N., Bak, P., Keim, D., Wrobel, S.: Visual Analytics of Movement. Springer, Heidelberg (2013). https://doi.org/10.1007/978-3-642-37583-5
3. Ankerst, M., Breunig, M.M., Kriegel, H., Sander, J.: OPTICS: ordering points to identify the clustering structure. In: SIGMOD, pp. 49–60 (1999)
4. Ayhan, S., Samet, H.: Aircraft trajectory prediction made easy with predictive analytics. In: Proceedings of the ACM SIGKDD 2016 (2016)
5. Ayhan, S., Samet, H.: Time series clustering of weather observations in predicting climb phase of aircraft trajectories. In: Proceedings of the IWCTS 2016 (2016)
6. Cheng, T., Cui, D., Cheng, P.: Data mining for air traffic flow forecasting: a hybrid model of neural network and statistical analysis. In: Proceedings of the ITSC 2003 (2003)
7. Ciccio, C.D., var der Aa, H., Cabanillas, C., et al.: Detecting flight trajectory anomalies and predicting diversions in freight transportation. Decis. Support. Syst. **88**, 1–17 (2016)

8. Coppenbarger, R.: En route climb trajectory prediction enhancement using airplane flight-planning information. American Institute of Aeronautics and Astronautics (AIAA-99-4147) (1999)
9. Deng, Z., Hu, Y., Zhu, M., Huang, X., Du, B.: A scalable and fast OPTICS for clustering trajectory big data. Clust. Comput. **18**(2), 549–562 (2015)
10. Enea, G., Poretta, M.: A comparison of 4D-trajectory operations envisioned for Nextgen and SESAR. In: Proceedings of the ICAS 2012 (2012)
11. Ester, M., Kriegel, H., Sander, J., Xu, X.: A density-based algorithm for discovering clusters in large spatial databases with noise. In: KDD, pp. 226–231 (1996)
12. Fablec, Y.L., Alliot, J.: Using neural networks to predict aircraft trajectories. In: Proceedings of the ICIS 1999 (1999)
13. Fan, Q., Zhang, D., Wu, H., Tan, K.: A general and parallel platform for mining co-movement patterns over large-scale trajectories. PVLDB **10**(4), 313–324 (2016)
14. Georgiou, H., et al.: Moving objects analytics: survey on future location & trajectory prediction methods (2018)
15. Georgiou, H., Pelekis, N., Sideridis, S., Scarlatti, D., Theodoridis, Y.: Semantic-aware aircraft trajectory prediction using flight plans. Int. J. Data Sci. Anal., 1–14 (2019). https://doi.org/10.1007/s41060-019-00182-4
16. Gong, C., McNally, D.: A methodology for automated trajectory prediction analysis (2004)
17. Hu, C., Kang, X., Luo, N., Zhao, Q.: Parallel clustering of big data of spatio-temporal trajectory. In: ICNC, pp. 769–774 (2015)
18. Jeung, H., Yiu, M.L., Zhou, X., Jensen, C.S., Shen, H.T.: Discovery of convoys in trajectory databases. PVLDB **1**(1), 1068–1080 (2008)
19. Kalnis, P., Mamoulis, N., Bakiras, S.: On discovering moving clusters in spatio-temporal data. In: Bauzer Medeiros, C., Egenhofer, M.J., Bertino, E. (eds.) SSTD 2005. LNCS, vol. 3633, pp. 364–381. Springer, Heidelberg (2005). https://doi.org/10.1007/11535331_21
20. Laube, P., Imfeld, S., Weibel, R.: Discovering relative motion patterns in groups of moving point objects. IJGIS **19**(6), 639–668 (2005)
21. Lee, J., Han, J., Whang, K.: Trajectory clustering: a partition-and-group framework. In: SIGMOD, pp. 593–604 (2007)
22. de Leege, A., Paassen, M.V., Mulder, M.: A machine learning approach to trajectory prediction. In: Proceedings of the AIAA GNC 2013 (2013)
23. Li, Y., Bailey, J., Kulik, L.: Efficient mining of platoon patterns in trajectory databases. Data Knowl. Eng. **100**, 167–187 (2015)
24. Li, Z., Ding, B., Han, J., Kays, R.: Swarm: mining relaxed temporal moving object clusters. PVLDB **3**(1), 723–734 (2010)
25. Marz, N., Warren, J.: Big Data: Principles and Best Practices of Scalable Real-Time Data Systems. Manning Publications Co., New York (2015)
26. Nanni, M., Pedreschi, D.: Time-focused clustering of trajectories of moving objects. J. Intell. Inf. Syst. **27**(3), 267–289 (2006)
27. Orakzai, F., Calders, T., Pedersen, T.B.: Distributed convoy pattern mining. In: IEEE MDM, pp. 122–131 (2016)
28. Orakzai, F., Calders, T., Pedersen, T.B.: k/2-hop: fast mining of convoy patterns with effective pruning. PVLDB **12**(9), 948–960 (2019)
29. Panagiotakis, C., Pelekis, N., Kopanakis, I.: Trajectory voting and classification based on spatiotemporal similarity in moving object databases. In: Adams, N.M., Robardet, C., Siebes, A., Boulicaut, J.-F. (eds.) IDA 2009. LNCS, vol. 5772, pp. 131–142. Springer, Heidelberg (2009). https://doi.org/10.1007/978-3-642-03915-7_12

30. Pelekis, N., Theodoridis, Y.: Mobility Data Management and Exploration. Springer, New York (2014). https://doi.org/10.1007/978-1-4939-0392-4
31. Pelekis, N., Tampakis, P., Vodas, M., Doulkeridis, C., Theodoridis, Y.: On temporal-constrained sub-trajectory cluster analysis. Data Min. Knowl. Discov. **31**(5), 1294–1330 (2017)
32. Pelekis, N., Tampakis, P., Vodas, M., Panagiotakis, C., Theodoridis, Y.: In-DBMS sampling-based sub-trajectory clustering. In: EDBT, pp. 632–643 (2017)
33. Ray, C., Dréo, R., Camossi, E., Jousselme, A.L.: Heterogeneous integrated dataset for maritime intelligence, surveillance, and reconnaissance (2018). https://doi.org/10.5281/zenodo.1167595
34. Seki, K., Jinno, R., Uehara, K.: Parallel distributed trajectory pattern mining using hierarchical grid with mapreduce. IJGHPC **5**(4), 79–96 (2013)
35. Tampakis, P., Pelekis, N., Andrienko, N.V., Andrienko, G.L., Fuchs, G., Theodoridis, Y.: Time-aware sub-trajectory clustering in Hermes@ PostgreSQL. In: ICDE, pp. 1581–1584 (2018)
36. Tampakis, P., Pelekis, N., Doulkeridis, C., Theodoridis, Y.: Scalable distributed subtrajectory clustering (2019). http://arxiv.org/abs/1906.06956
37. Tang, L.A., et al.: On discovery of traveling companions from streaming trajectories. In: ICDE, pp. 186–197 (2012)
38. Tao, Y., Faloutsos, C., Papadias, D., Liu, B.: Prediction and indexing of moving objects with unknown motion patterns. In: Proceedings of the ACM SIGMOD 2004 (2004)
39. Thipphavong, D., Schultz, C., et al.: Adaptive algorithm to improve trajectory prediction accuracy of climbing aircraft. J. Guid. Control. Dyn. (JGCD) **36**(1), 15–24 (2013)
40. Trasarti, R., Guidotti, R., Monreale, A., Giannotti, F.: MyWay: location prediction via mobility profiling. Inf. Syst. **64**, 350–367 (2017)
41. Vieira, M.R., Bakalov, P., Tsotras, V.J.: On-line discovery of flock patterns in spatio-temporal data. In: ACM SIGSPATIAL, pp. 286–295 (2009)
42. Zheng, K., Zheng, Y., Yuan, N.J., Shang, S.: On discovery of gathering patterns from trajectories. In: ICDE, pp. 242–253 (2013)
43. Zheng, Y.: Trajectory data mining: an overview. Trans. Intell. Syst. Technol. **6**(3), 1–41 (2015)
44. Zheng, Y.: Trajectory data mining: an overview. ACM TIST **6**(3), 29:1–29:41 (2015)

EvolvingClusters: Online Discovery of Group Patterns in Enriched Maritime Data

George S. Theodoropoulos, Andreas Tritsarolis(✉), and Yannis Theodoridis

Data Science Laboratory, University of Piraeus, Piraeus, Greece
{gstheo,andrewt,ytheod}@unipi.gr
http://www.datastories.org/

Abstract. In this paper, we propose a novel unified online group pattern mining algorithm, *EvolvingClusters*, that aims to enrich geospatial data through the mapping of their group behaviour. Specifically, *EvolvingClusters* is used to discover collective movement behaviour (like flocks and convoys) by monitoring the activity of multiple clusters through time and space. We evaluate the aforementioned algorithm using a real-world marine traffic dataset consisting of vessels' movement in Brest Bay, France. Our study demonstrates the efficiency and effectiveness of the proposed algorithm as well as its value towards a semantic enrichment tool that can be used to observe and categorize the behaviour of multiple moving objects in real time.

Keywords: Big data · Data analytics · Maritime Intelligence · Collective movement behaviour · Group patterns · Flocks · Convoys · Semantic enrichment

1 Introduction

Mobility Data Analytics [2,9,21] is an ever growing branch of the general spectrum of Data Analytics. GPS-enabled mobile phones, cars, airplanes, and vessels are the most common data sources broadcasting volumes of location information. Using them as-is (i.e., in their "raw" form), offer us limited usefulness; however, with proper processing (cleansing, transformation, enrichment etc.) and analysis (pattern discovery), the vast amount of available data can produce some very interesting and insightful stories. The outcome of data analytics over mobility data is of great interest to researchers and practitioners of the field.

More specifically, in the field of semantic enrichment, behavioural clustering can provide a concise and meaningful base that can be of value to multiple mining methodologies. Classification with the use of artificial neural networks for example, is a process that requires vast amounts of data, computational resources and time. Using a behavioural clustering technique like EvolvingClusters can be very beneficial, especially with respect to time and resources, since the classifier will be able to train on a smaller set of objects that belong to multiple different

K. Tserpes et al. (Eds.): MASTER 2019, LNAI 11889, pp. 50–65, 2020.
https://doi.org/10.1007/978-3-030-38081-6_5

clusters instead of the full dataset that might contain objects with a lot of similarities.

This paper focuses on mobility data analytics over maritime traffic data. In particular, our purpose is to evaluate group movement behaviour at sea (e.g. flocks, convoys) over enriched trajectories of vessels.

The contributions of our work are summarized in the following lines:

- We enrich vessel movement data with annotations regarding their closeness to ports, etc.
- We design and evaluate a unified group behaviour discovery algorithm able to simulate existing pattern discovery methods, such as flocks and convoys.
- We evaluate the above over a large-volume real-world maritime trajectory dataset [22].

Our paper is structured as follows: In Sect. 2, we present background knowledge and related work. In Sect. 3, we provide our problem formulation and discuss what is special about maritime data. In Sect. 4, we present our *Evolving Clusters* algorithm for unified group pattern mining. In Sect. 5, we discuss preliminary experimental results. Section 6 concludes the paper, also giving hints for future work.

2 Background Knowledge and Related Work

The field of trajectory data mining [27] is rich in methods capturing collective movement of objects, i.e. sets of objects moving close to each other for a certain time period.

Flocks [4,10,25] take into account the spatial proximity and the direction of moving objects. For a flock pattern to be discovered, a minimal number of trajectories that satisfy such constraints are required. Formally, a flock valid during a time interval I, where I spans for at least k successive timepoints, consists of at least m objects, such that for every timepoint in I, there is a disk of radius r that contains all those entities. Technically, a flock discovery algorithm is tuned by three parameters: k (the minimum number of successive timepoints), m (the minimum number of neighboring objects), and r (the radius that defines the neighborhood). Companion [24] and Gathering [26] are two flock variations, focusing on online/streaming applications.

A *convoy* [12,13,20] is a group of objects consisting of at least m objects that are density-connected with respect to a density-reachability distance threshold e, during at least k consecutive timepoints. Specifically, assuming the partitioning of the database of the objects' locations with respect to a discretization of the time dimension, a snapshot S_i (i.e., the set of objects and their locations that exist at time t_i), is clustered using a typical density-based spatial clustering algorithm like DBSCAN [7], to identify dense groups of objects in S_i that are close to each other and the density of the group meets the density constraints of the clustering algorithm, i.e. the minimum number of objects in an object's neighborhood, $MinPts$, and the maximum distance for two objects to be directly density-reachable, e, according to DBSCAN's parameters. Technically, a convoy discovery algorithm is tuned by three parameters: k, m (as defined in flocks above), and e. Compared to flocks, convoys actually differ in that the circular neighborhood is replaced by the notion of density-connection. Convoy variations include *groups* [17] and *evolving groups* [15].

A *swarm* [19] is a collection of moving objects with cardinality of at least m, that are part of the same density-based cluster, defined by a reachability distance threshold e, for at least k (not necessarily consecutive) timepoints. Moreover, comparing the clusters themselves, the population is not required to remain unchanged but at least one cluster containing all objects should be discovered. Note that the trajectory of each object in-between these timepoints, is not under any constraint. Technically, a swarm discovery algorithm is tuned by the same three parameters, k, m, and e, as in the cases of moving clusters and convoys above, with the main difference being that swarms do not require the set of at least k timestamps to be consecutive.

Further related work includes the following. A *moving cluster* [14] is a sequence of clusters $c_1, \ldots, c_k$, such that for each timestamp t_i, clusters c_i and c_{i+1} share a sufficient number of common objects. Intuitively, if the two spatial clusters at two consecutive snapshots have a large percentage of common objects then they are considered a moving cluster between these two timestamps. A *moving micro cluster* [18] is a group of objects that are not only close to each other at the current time, but they are also expected to move together in the near future; techniques for maintaining clusters of moving objects by considering the clusters of the current and near-future positions are proposed in [11]; [6] presents a taxonomy/classification of movement patterns along a set of dimensions that reveal their behavior (and commonalities); [3] demonstrates the shortcomings of the Jaccard (J) measure when it is used for assessing the significance of co-occurrences among spatiotemporal instances with highly different spatiotemporal evolution characteristics and presents two extended novel measures (J^+ and J^*) that address the problems linked to the J measure; [5] studies a regional semantic trajectory pattern mining problem, aiming at identifying all the regional sequential patterns in semantic trajectories.

Most related to our work, [16] defines various mobility behaviors around the idea of the flock pattern; in particular, the Relative Motion (REMO) model and a respective language are proposed in order to express a number of collective mobility patterns under a unified representation. [23] proposes, among others, gpattern and crosspattern, two generic query operators implemented and validated in the *Secondo MOD system* [1], which express groups of moving objects that follow similar motion and mutually interact together, respectively (mobility behaviors, such as flocking, convergence, and leadership can be simulated through these operators).

With respect to related work, our method handles closeness of moving objects in a unified way under a graph-based approach, being able to simulate the most popular patterns (i.e. flocks and convoys) in an online mode.

3 Problem Formulation

An informal *group pattern* definition could be: "a large enough amount of objects moving along paths close to each other for a certain time". These objects could vary from animals (e.g. wolves, birds, lions, etc.) to human transportation means (e.g. cars, airplanes and vessels). Discovering these patterns can give us an insight regarding the behavior of these moving objects, for instance on hunting (wolves, lions), migration (birds), traffic monitoring (cars) and fishing pressure (fishing vessels). In this paper we aim at handling group pattern discovery in a uniform way, where "closeness" is formulated in graph-based terminology.

3.1 Problem Definition

Definition 1. *(Evolving Cluster). Given: a set T of moving objects, where the trajectory of each object consists of r pairs (p_i, t_i), a minimum cardinality threshold c, a maximum distance threshold θ, and a minimum time duration threshold d, an Evolving Cluster $\langle C, t_{start}, t_{end}, tp\rangle$ is a subset $C \in T$ of the moving objects' population, $|C| \geq c$, which appeared at time point t_{start} and remained alive until time point t_{end} (with $t_{end} - t_{start} \geq d$) during the lifetime $[t_{start}, t_{end}]$ of which the participating moving objects were spatially connected with respect to distance θ and cluster type tp.*

The term "spatially connected" is used on purpose in the above definition, since the structure of our method accounts for a number of different clustering methodologies. In this study, we use both spherical and density-based clustering in order to mine flock and convoy-like patterns, respectively. In particular, for each time point, let us consider the mapping of the points of the moving objects' trajectories (that are active at that time point) in a connectivity graph $G(V, E)$, where vertex $v \in V$ represents a point and edge $e \in E$ represents a pair of points if and only if their distance is less than the given threshold θ; *Cliques* in this graph correspond to spherical-like clusters whereas *Maximal Connected Subgraphs (MCS)* in this graph correspond to density-connected clusters. Cliques (maximal connected subgraphs) that remain alive for an adequate period of time are evolving clusters, according to the above definition, resembling flock (convoy, respectively) patterns. (Please note that in the discussion that follows, when we use the term Cliques we refer to maximal Cliques.) This concept is better illustrated in Fig. 1.

According to Fig. 1, sets $C_1 = \{a, b, c, d\}$ and $C_2 = \{a, b, c, d, e, f\}$ form a Clique and an MCS, that remain active for three time points $t_1, \ldots, t_3$, while $C_3 = \{a, b, c\}$ and $C_4 = \{d, e, f\}$ form a Clique and an MCS, that remain active during all four time points $t_1, \ldots, t_4$. Assuming thresholds e.g. $c = 3$ and $d = 3$, we have discovered three *Evolving Clusters*, the spherical-like $\langle C_1, t_1, t_3, 1\rangle$ and $\langle C_3, t_1, t_4, 1\rangle$, and the density-connected $\langle C_2, t_1, t_3, 2\rangle$ and $\langle C_4, t_1, t_4, 2\rangle$, where cluster type 1(2) corresponds to Clique (MCS, respectively). This example illustrates that two evolving clusters can be overlapping with respect to their population.

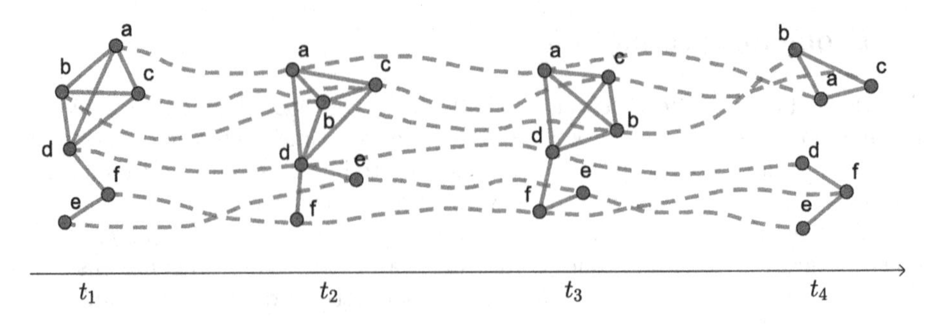

Fig. 1. An example of six objects moving at four consecutive time points and the respective connectivity graphs.

3.2 What Is Special About Maritime Data

It is well known that sensor-based information is sensitive to errors due to device malfunctioning. Therefore, a necessary step before performing data analytics tasks is that of pre-processing. A necessary clarification is that since the vessels' locations are recorded in angular (lat/lon) coordinates, we use the Haversine formula as it takes in account the data points' geodesic properties.

In general, pre-processing of GPS-based location data includes data cleansing (noise elimination, location smoothing, etc.) as well as data transformation tasks necessary for the analysis that will follow (fixed rate resampling, trajectory segmentation, etc.) [21]. A typical data preprocessing workflow consists of the following steps:

1. Data Cleansing:
    a. Remove time-based duplicate records;
    b. Remove position-based outliers (i.e. invalid speed, acceleration, etc.);
2. Data transformation
    a. Create Trips from vessels' locations;
    b. (Optional) Perform fixed-rate resampling on Trips;

In particular for Step 2a and in order to organize vessels' locations in trips, a popular approach (in case the ports are given as points instead of polygons) is to create a circle with radius ρ around each port's location in order to approximate their geometry and then, detect port entry and exit points for each vessel trajectory (Spatial-based Segmentation).

Then, for each produced segment, we may detect pairs of points with temporal difference greater than a given threshold (Temporal-based Segmentation). These pairs signify the transition from the current to the next *Trip*.

The segmentation due to the above steps, may result in a very low number of points. Because they do not offer any significant information, we decide to filter out these particular *Trips* (in particular, those consisting of less than 3 points).

Depending on their connection with ports, vessels' trips can be classified in 4 classes:

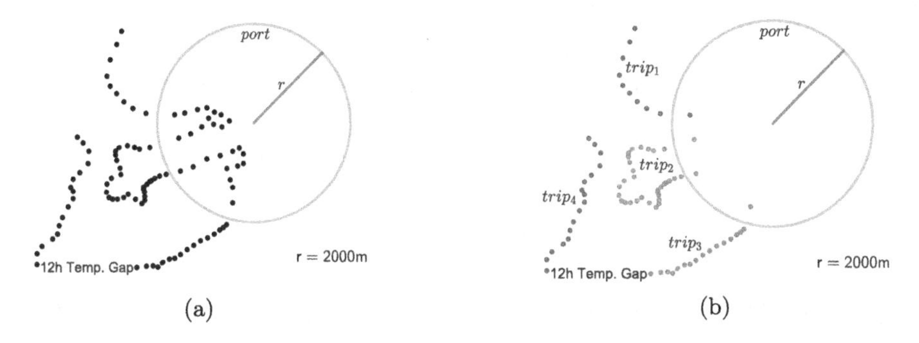

Fig. 2. A sample trajectory: (a) before; and (b) after trajectory segmentation into trips.

- Class C1 — trips that start and end at a port;
- Class C2 — trips that start at a port and end at open sea;
- Class C3 — trips that start at open sea and end at a port;
- Class C4 — trips that start and end at open sea;

The aforementioned methodology is illustrated in Fig. 2, where the raw location information is compared with a port's location, hence a vessel trajectory is segmented into trips of Class C1 (e.g. $trip_2$ in Fig. 2(b), Class C2 ($trip_3$), Class C3 ($trip_1$), and Class C4 ($trip_4$).

Given that a vessel traffic dataset consists of GPS points that are sampled whenever the captain of each vessel enables the AIS transmitter, it is obvious that there is no form of consistency regarding the time intervals between points. For example, it is easily observable that a vessel is highly likely to stop transmitting for a considerable amount of time if that vessel is inactive, e.g being stationary on a port. As a result, while also keeping in mind that several techniques used for future location prediction as well as group pattern mining need or benefit substantially by a stable rate of sampling and, by extension, a temporal alignment, a fixed-rate resampling technique [8] is used to achieve the consistency needed; see Fig. 3 for an illustration of the above discussion.

4 The EvolvingClusters Algorithm

In this section we present an algorithm, called *EvolvingClusters*, in order to detect and extract group patterns from raw GPS data points. This algorithm is fully modular, mining clusters with respect to the spatial clustering restrictions stated in the previous sections and then by applying the temporal restrictions, can fetch different types of group patterns simultaneously (in our case, *Cliques* and *MCS*). Due to the fact that we only compare our pattern history with the current time-slice, the algorithm can be connected to a data stream, thus having an online nature.

Algorithm 1 presents the algorithm's corpus. In particular it discovers evolving clusters in a trajectory dataset D, where moving objects' locations arrive

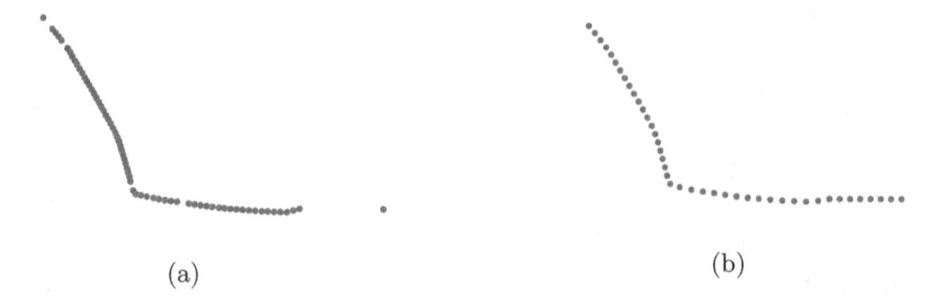

(a) (b)

Fig. 3. A sample vessel trip: (a) before; and (b) after resampling (with 1 min. fixed sampling rate).

Algorithm 1. EVOLVINGCLUSTERS. An online algorithm capable of mining the Group Patterns as discussed in the previous subsections

---

Input: A Dataset $D = \{T_1, T_2, \ldots, T_n\}$ of Time-slices T_i consisting of objects' timestamped locations (p_j, t_i), Distance Threshold θ, Time Threshold t, Cardinality Threshold c

Output: A list of all the mined patterns $MinedPatterns$

```
1  ActivePatterns, ClosedPatterns ← []
2  for Time-slice T in D do
3      C_Cliques, C_MCS ← GeospatialClustering(T, θ, c)
4      for CurrentClusters in {C_Cliques, C_MCS} do
5          if ActivePatterns == ∅ then
6              ActivePatterns ← CurrentClusters
7          else if CurrentClusters == ∅ then
8              ClosedPatterns ← {ActivePattern ∈ ActivePatterns :
                 ActivePattern.end − ActivePattern.start ≥ t}
9          else
10             ActivePatterns, InactivePatterns ←
                 FindPatterns(CurrentClusters, ActivePatterns, θ)
11             ClosedPatterns ← {InactivePattern ∈ InactivePatterns :
                 InactivePattern.end − InactivePattern.start ≥ t}
12         end
13         output
14           {Pattern ∈ ActivePatterns : Pattern.end − Pattern.start ≥ t}
15     end
16 end
```

---

at predefined timepoints (e.g. every 60 sec.) or, in other words, at a fixed (and aligned amongst all objects) sampling rate.

In the following paragraphs we provide a thorough explanation regarding its operation. Algorithm 1 is responsible of using the results provided by Algorithm 3 in a sequential manner. Essentially Algorithm 1 uses the results of Algorithm 2 and decides if the available data in the form of *ActivePatterns* (patterns previously mined) and *CurrentClusters* (clusters formed based on the location of moving objects at the current time-slice) are eligible to be used as input to Algorithm 3. If not (either set is empty), the algorithm either moves the clusters currently active to the *ActivePatterns* set (if $ActivePatterns = \emptyset$) or moves all the patterns that satisfy the thresholds given from *ActivePatterns* to *ClosedPatterns* (if $CurrentClusters = \emptyset$).

Algorithm 2. GEOSPATIAL CLUSTERING. Clusters GPS Points given a Time-slice

Input: Time-slice $T = \{p_1, p_2, \ldots, p_n\}$ of coordinate points, Distance threshold θ, Cardinality threshold c

Output: Clusters of the Time-slice's Points $C_{Cliques}, C_{MCS}$

1 $DistanceMatrix \leftarrow PairwiseDistance(T,\ metric = \text{"}Haversine\ Distance\text{"})$
2 $Pairs \leftarrow \{(p_i, p_j) : DistanceMatrix(p_i, p_j) < \theta\}$
3 $G \leftarrow \text{Graph}(edges = pairs)$
4 $C_{Cliques} \leftarrow \{C \in G.Cliques() : |C| \geq c\}$
5 $C_{MCS} \leftarrow \{C \in G.MaximalConnectedSubgraphs() : |C| \geq c\}$
6 **return** $C_{Cliques}, C_{MCS}$

Algorithm 3 takes all the following *cases* into consideration: (for pattern C_{t_i} at time t_i and $C_{t_{i+1}}$ at time t_{i+1})

1. The patterns are identical ($C_{t_i} = C_{t_{i+1}}$)
2. The patterns have no common objects ($C_{t_i} \cap C_{t_{i+1}} = \emptyset$)
3. The pattern C_{t_i} is a subset of $C_{t_{i+1}}$ ($C_{t_i} \subset C_{t_{i+1}}$)
4. The pattern $C_{t_{i+1}}$ is a subset of C_{t_i} ($C_{t_{i+1}} \subset C_{t_i}$)
5. The patterns contain some common objects ($C_{t_i} \cap C_{t_{i+1}} \neq \emptyset$, $C_{t_i} \cap C_{t_{i+1}} \subset C_{t_i}, C_{t_{i+1}}$)

Therefore, the algorithm operates as follows:

- For every pair of consecutive (with respect to time) pattens, if the cardinality of their intersection is greater than c, add it to the *ActivePatterns* set (lines: 4–7).
- For every pattern in $C_{t_{i+1}}$, if the list of its intersections with all of the patterns in C_{t_i} doesn't contain the pattern, add it to the *ActivePatterns* set as a new pattern (lines: 8–9).
- For every pattern in C_{t_i}, if it is not part of the *ActivePatterns* set, add it to the *InactivePatterns* set (line: 11).
- Replace each group of duplicate patterns in the *ActivePatterns* set, with a single record of each pattern and substitute its starting and ending timestamps with the oldest starting and newest ending timestamps available in the duplicate group (lines: 12–17).

We observe that in all *cases* the pattern that ought to be maintained through time is the intersection of C_{t_i} and $C_{t_{i+1}}$. *Cases 2 and 3* require some extra attention. Regarding *case 2*, the intersection is an empty set. As a result, $C_{t_{i+1}}$ should be maintained and added to the *ActivePatterns* set. *Case 3* dictates that both the new superset and the previous pattern should be maintained since they both exist at the same time as part of $C_{t_{i+1}}$.

Algorithm 3. FINDPATTERNS. Compares the current with the closed clusters in order to determine their evolution

Input: Consecutive datasets D_{left}, D_{right}, Cardinality threshold c
Output: Mined patterns $ActivePatterns, InactivePatterns$

```
1  ActivePatterns ← []
2  for Pattern P_R in D_right do
3      IntersectionList ← {}
4      for Pattern P_L in D_left do
5          if |P_R ∩ P_L| ≥ c then
6              ActivePatterns.append([[P_R ∩ P_L, P_L.start, P_R.end]])
7      end
8      if IntersectionList = ∅ then
9          ActivePatterns ← P_R
10 end
11 InactivePatterns ← {pattern ∈ P_L : pattern ∉ ActivePatterns}
12 for Pattern P_active in ActivePatterns do
13     DuplicatePatterns ← [pattern_A, pattern_B ∈ P_active : (pattern_A =
         pattern_B) ∧ (pattern_A ≠ pattern_B)]
14     if |DuplicatePatterns| ≠ 0 then
15         P_active.start ← min(DuplicatePatterns.start)
16         P_active.end ← max(DuplicatePatterns.end)
17 end
18 return ActivePatterns, InactivePatterns
```

Based on Fig. 1 the proposed algorithm for $c = 3, t = 3$ would mine the patterns $C_1 = \{a, b, c, d\}$, $C_2 = \{a, b, c, d, e, f\}$, $C_3 = \{a, b, c\}$ and $C_4 = \{d, e, f\}$ as follows:

- t_1: Clique $\langle C_1, t_1, t_1, 1\rangle$ and MCS $\langle C_2, t_1, t_1, 2\rangle$ mined (Output: $\emptyset$);
- t_2: Clique $\langle C_1, t_1, t_2, 1\rangle$ and MCS $\langle C_2, t_1, t_2, 2\rangle$ mined (Output: $\emptyset$);
- t_3: Clique $\langle C_1, t_1, t_3, 1\rangle$ and MCS $\langle C_2, t_1, t_3, 2\rangle$ mined (Output: $\{C_1, C_2\}$);
- t_4: Clique $\langle C_3, t_1, t_4, 1\rangle$ and MCS $\langle C_4, t_1, t_4, 2\rangle$ mined (Output: $\{C_3, C_4\}$).

For timestamps t_1 through t_3, C_1 and C_2 are mined and maintained. During timestamp t_4, two new patterns are found (C_3 and C_4), however both new patterns are present during t_3 as *subsets* of C_1 and C_2 respectively. Thus they get to keep the starting timestamp of their respective *past* supersets.

5 Experimental Study

In this section we prepare the dataset that the algorithm will be tested on and present some preliminary results regarding its effectiveness.

5.1 Dataset Preparation

In our study, we use a publicly available maritime dataset, called Heterogeneous Integrated Dataset for Maritime Intelligence, Surveillance, and Reconnaissance [22], which contains information on maritime traffic in France. The dataset ranges in time and space as follows:

- temporal range: October 1st, 2015 to March 31st, 2016 (6 months);
- spatial range: latitude in [45.00, 51.00], longitude in [-10.00, 0.00] (Celtic sea, the Channel and Bay of Biscay).

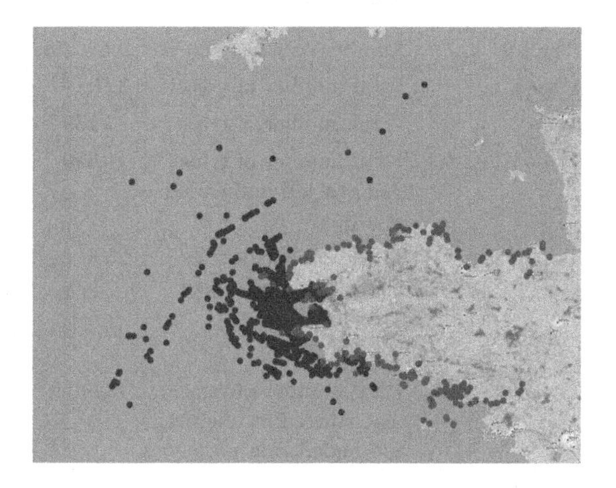

Fig. 4. A snapshot from the Brest dataset: sample of AIS positions on March 1st, 2016 (blue dots) and ports of interest (red dots). (Color figure online)

A map visualization of (a part of) the dataset is illustrated in Fig. 4. The original dataset contains three classes of information: *vessel-dynamic* (i.e., related to the vessels' movement), *vessel-static* (i.e., related to the vessels' identity), and *geo-related* data (locations of ports, environmental information, etc.). For the purposes of our study, we exploit on the entire vessel-dynamic and vessel-static information available while from the third class we are only interested in the locations of ports, information which is essential for the analysis we design to perform. In particular:

- The *vessel-dynamic* data contains approximately 19 million records. Each record corresponds to an AIS signal and includes the vessel identity (mmsi), its position (lon, lat), the timestamp this position was recorded (ts) as well as other mobility-related information provided by vessel's sensors (speed, course, heading, etc.).

- The *vessel-static* data contains information about vessel registration, such as the vessel's identity (mmsi), radio frequency (call sign), name, type, size, etc.
- The *geo-related* data we used in our study contains 222 records; each record corresponds to a port along with its name and location (point geometry).

Due to step 2a (recall the preprocessing steps of Sect. 3.2), with port radius set at 2 km ($\approx$1.08 n.m.) and temporal threshold at 12 h, trajectory segmentation yields 9,545,789 data points organized in 24,159 trips from 3,279 vessels (segments with very few data points - i.e. less than 3 - are discarded).

Table 1. Statistics of the dataset after the pre-processing step.

#Records	Number of AIS Records	9,545,789
#Vessels	Total number of vessels	3279
#Trips	Total number of trips	24,159
#Trips Class C1	Total number of trips that started and ended at a port	11,690
#Trips Class C2	Total number of trips that started at a port and ended	2580
	at open sea	
#Trips Class C3	Total number of trips that started at open sea and ended at a port	1849
#Trips Class C4	Total number of trips that started and ended at open sea	8040

5.2 Preliminary Results

Having processed our dataset using the methodology presented in Sect. 3.2, we tested our algorithm on a wide range of values for each parameter, namely:

- Cardinality Threshold (c): $3, 5, 8, 12$. Default: 5 vessels
- Temporal Threshold (t): $15, 30, 45, 60$. Default: 15 min
- Distance Threshold (θ): $0.25, 0.5, 0.75, 1, 1.25$. Default: 1 nautical mile

Figure 5 illustrates the average percentage of trip classes C1–C4 in the mined group patterns (using the default parameters). In either pattern type, we observe that C1 is the most dominant class (having more than 60% participation), which is reasonable since the same participation appears – more or less – at trip level (see Table 1). On the other hand, C4 presents an interesting behaviour: although its percentage at trip level is about 30%, this percentage falls down to 13% within cliques and 7.7% within MCSs. Comparing the two pattern types (cliques and MCSs) with each other, we observe that cliques appear to be formed more frequently than MCSs when we focus on C3 or C4 while the opposite is the case when we focus on C1 or C2. These findings may trigger domain experts to take a deeper look and reach insightful conclusions.

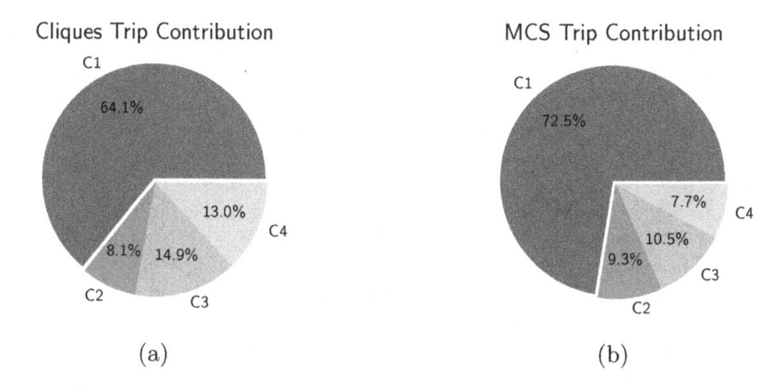

Fig. 5. Trip contribution on mined (a) Cliques (b) MCS.

Figure 6 illustrates the change of *average distance travelled* (*#group patterns*, respectively) with respect to one of the algorithm's parameters, while the others are fixed to their respective default values. It can be observed that as c increases, both types of group patterns decline both in their respective average distance travelled and their cardinality (Fig. 6a and b, respectively), while on the other hand, as θ increases, the opposite can be seen (Fig. 6c and d, respectively). Moreover, as t increases, we observe a steady rise in the average distance travelled for both pattern types but at the cost of having fewer patterns (Fig. 6e and f, respectively). Furthermore, it is shown that *Cliques* are quite sensitive with respect to their thresholds while *MCS* as less sensitive, showing a more steady growth/decline (Fig. 6b, d and f). Last but not least, as illustrated by Fig. 6a, c and e, a linear-like correlation can be observed between the thresholds c, t, θ and the average distance travelled.

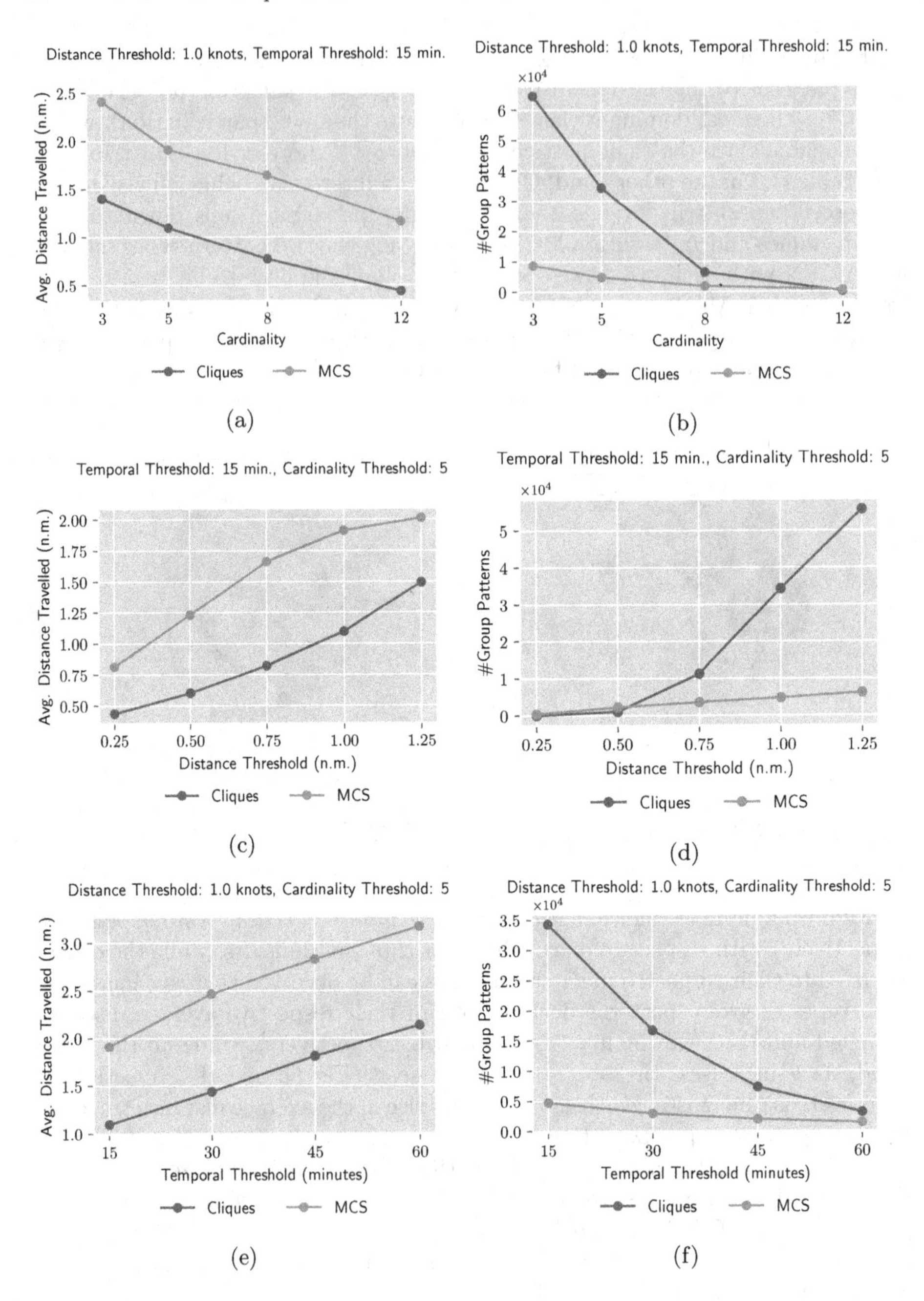

Fig. 6. Statistics on mined group patterns: cardinality, distance and temporal threshold vs. (a, c, e) avg. distance travelled and (b, d, f) #group patterns.

6 Conclusions and Future Work

In this paper, we proposed a unified online group pattern mining algorithm, called EvolvingClusters, which is used to discover collective movement behaviour (like flocks and convoys) by monitoring the activity of multiple clusters through time and space. The algorithm is graph-based in the sense that it maintains evolving *Cliques* and *Maximal Connected Subgraphs (MCS)*, thus simulating spherical and density-based evolving clusters. Our study on a large real-world maritime traffic dataset demonstrates the efficiency and effectiveness of the proposed algorithm. The results show that our method is capable of detecting a large amount of group patterns in the given dataset. Thus, based on the potential applications, some of which were mentioned above, as well as the quality of the results produced, we believe that *EvolvingClusters* can be a valuable tool for researchers and practitioners alike.

In the near future we aim to test and evaluate *EvolvingClusters* against other state-of-the-art techniques, using other types of mobility data, such as aviation and public transportation data. Based on our assumptions, the algorithm should function at the same quality level no matter the data type used, since its approach does not make use of any other apriori form of knowledge like road grids or hot-paths. Another set of experiments that we would like to conduct in the near future is using data with different sampling rates as input for *EvolvingClusters*. If the results appear to be in the same quality level with those produced from a dataset with a much higher sampling rate as input, we would be certain that the value of the algorithm is not tied to the sampling rate of the given data. Our long-term plans involve around the creation of a framework that will use the information that is mined using *EvolvingClusters* to classify moving objects into different classes based on their behaviour. By extracting as much information as possible from the available data and combining a well trained classifier with a well defined set of groups with similar behaviour, we want to create a system able to model and – if possible – predict a set of suspicious activities that might consist a violation of law or a possible criminal activity.

Acknowledgments. This work was partially supported by the Greek Ministry of Development and Investment, General Secretariat of Research and Technology, under the Operational Programme Competitiveness, Entrepreneurship and Innovation 2014–2020 (grant T1EDK-03268, i4Sea).

References

1. de Almeida, V.T., Güting, R.H., Behr, T.: Querying moving objects in SECONDO. In: MDM, p. 47. IEEE Computer Society (2006)
2. Andrienko, G.L., Andrienko, N.V., Bak, P., Keim, D.A., Wrobel, S.: Visual Analytics of Movement. Springer, Heidelberg (2013)
3. Aydin, B., Küçük, A., Angryk, R.A., Martens, P.C.: Measuring the significance of spatiotemporal co-occurrences. ACM Trans. Spat. Algorithms Syst. **3**(3), 9:1–9:35 (2017)

4. Benkert, M., Gudmundsson, J., Hübner, F., Wolle, T.: Reporting flock patterns. Comput. Geom. **41**(3), 111–125 (2008)
5. Choi, D., Pei, J., Heinis, T.: Efficient mining of regional movement patterns in semantic trajectories. PVLDB **10**(13), 2073–2084 (2017)
6. Dodge, S., Weibel, R., Lautenschütz, A.: Towards a taxonomy of movement patterns. Inf. Vis. **7**(3–4), 240–252 (2008)
7. Ester, M., Kriegel, H., Sander, J., Xu, X.: A density-based algorithm for discovering clusters in large spatial databases with noise. In: KDD, pp. 226–231. AAAI Press (1996)
8. Georgiou, H.: 1-pass fixed-rate linear resampler in Matlab/Octave (2017)
9. Giannotti, F., Pedreschi, D. (eds.): Mobility, Data Mining and Privacy - Geographic Knowledge Discovery. Springer (2008) https://doi.org/10.1007/978-3-540-75177-9
10. Gudmundsson, J., van Kreveld, M.J.: Computing longest duration flocks in trajectory data. In: GIS, pp. 35–42. ACM (2006)
11. Jensen, C.S., Lin, D., Ooi, B.C.: Continuous clustering of moving objects. IEEE Trans. Knowl. Data Eng. **19**(9), 1161–1174 (2007)
12. Jeung, H., Shen, H.T., Zhou, X.: Convoy queries in spatio-temporal databases. In: ICDE, pp. 1457–1459. IEEE Computer Society (2008)
13. Jeung, H., Yiu, M.L., Zhou, X., Jensen, C.S., Shen, H.T.: Discovery of convoys in trajectory databases. PVLDB **1**(1), 1068–1080 (2008)
14. Kalnis, P., Mamoulis, N., Bakiras, S.: On discovering moving clusters in spatio-temporal data. In: Bauzer Medeiros, C., Egenhofer, M.J., Bertino, E. (eds.) SSTD 2005. LNCS, vol. 3633, pp. 364–381. Springer, Heidelberg (2005). https://doi.org/10.1007/11535331_21
15. Lan, R., Yu, Y., Cao, L., Song, P., Wang, Y.: Discovering evolving moving object groups from massive-scale trajectory streams. In: MDM, pp. 256–265. IEEE Computer Society (2017)
16. Laube, P., Imfeld, S., Weibel, R.: Discovering relative motion patterns in groups of moving point objects. Int. J. Geogr. Inf. Sci. **19**(6), 639–668 (2005)
17. Li, X., Ceikute, V., Jensen, C.S., Tan, K.: Effective online group discovery in trajectory databases. IEEE Trans. Knowl. Data Eng. **25**(12), 2752–2766 (2013)
18. Li, Y., Han, J., Yang, J.: Clustering moving objects. In: KDD, pp. 617–622. ACM (2004)
19. Li, Z., Ding, B., Han, J., Kays, R.: Swarm: mining relaxed temporal moving object clusters. PVLDB **3**(1), 723–734 (2010)
20. Orakzai, F., Calders, T., Pedersen, T.B.: k/2-hop: fast mining of convoy patterns with effective pruning. PVLDB **12**(9), 948–960 (2019)
21. Pelekis, N., Theodoridis, Y.: Mobility Data Management and Exploration. Springer, New York (2014). https://doi.org/10.1007/978-1-4939-0392-4
22. Ray, C., Dreo, R., Camossi, E., Jousselme, A.L., Iphar, C.: Heterogeneous integrated dataset for maritime intelligence, surveillance, and reconnaissance. Data Brief **25**, 104140 (2019)
23. Sakr, M.A., Güting, R.H.: Group spatiotemporal pattern queries. GeoInformatica **18**(4), 699–746 (2014)
24. Tang, L.A., et al.: On discovery of traveling companions from streaming trajectories. In: ICDE, pp. 186–197. IEEE Computer Society (2012)
25. Vieira, M.R., Bakalov, P., Tsotras, V.J.: On-line discovery of flock patterns in spatio-temporal data. In: GIS, pp. 286–295. ACM (2009)

26. Zheng, K., Zheng, Y., Yuan, N.J., Shang, S., Zhou, X.: Online discovery of gathering patterns over trajectories. IEEE Trans. Knowl. Data Eng. **26**(8), 1974–1988 (2014)
27. Zheng, Y.: Trajectory data mining: an overview. ACM TIST **6**(3), 29:1–29:41 (2015)

Prospective Data Model and Distributed Query Processing for Mobile Sensing Data Streams

Mariem Brahem(✉), Karine Zeitouni, Laurent Yeh, and Hafsa El Hafyani

DAVID laboratory, University of Versailles Saint-Quentin-en-Yvelines, Paris Saclay University, Saint-Aubin, France
{mariem.brahem,karine.zeitouni,laurent.yeh,hafsa.el-hafyani}@uvsq.fr

Abstract. With the rapid advancements of sensor technologies and mobile computing, Mobile Crowd-Sensing (MCS) has emerged as a new paradigm to collect massive-scale rich trajectory data. Nomadic sensors empower people and objects with the capability of reporting and sharing observations on their state, their behavior and/or their surrounding environments. Processing and analyzing this continuously growing data raise several challenges due not only to their volume, their velocity, and their complexity but also to the gap between raw data samples and the desired application view in terms of correlation between observations and in terms of granularity. In this paper, we put forward a proposal that offers an abstract view of any spatio-temporal data series as well as their manipulation. Our approach allows to support this high-level logical view and provides efficient processing by mapping both the representation and the manipulation to an internal physical model. We explore an implementation within a distributed framework and envision the adaptation of data organization methods combining aggressive indexing and partitioning over time and space. The mapping from the logical view and the actual data storage will lead to revisiting the traditional database query rewriting and optimization techniques. This proposal is a first step in the objective of coping with the complexity, the imperfection of large data sizes in the MCS context.

Keywords: Spatio-temporal data modeling · Mobile crowd-sensing · Query processing

1 Introduction

The recent advances in sensing technologies and mobile computing have paved the way for the emergence of the Mobile Crowd-Sensing (MCS) [15,17] concept, leading to a continuous generation of large volume of rich trajectory data. More and more people rely on mobile devices (e.g., smartphone, tablets ...) and wearable sensors to share observations on their state, their behavior and/or their surrounding environments such as noise level, temperature or pollution conditions.

K. Tserpes et al. (Eds.): MASTER 2019, LNAI 11889, pp. 66–82, 2020.
https://doi.org/10.1007/978-3-030-38081-6_6

The growing scale of sensed data (Volume) coupled with real time sensor observations (Velocity) requires an effective model and efficient processing of complex spatio-temporal queries. In the past, processing large-scale data or high velocity data has been a bottleneck. Recently, big data analytics systems are becoming a de facto standard in massive data handling. While such systems fit well the large scale nature of sensed data, several issues related to the gap between raw data samples and the desired application view in term of correlation between observations occurring in different locations, or between different periods of time and spatial/temporal granularity (Variety) are still open. The heterogeneity and the diversity of sensor handsets from different manufacturers (with different sensitivities, time resolutions, and noise immunity) necessitate both an abstract data model, and an efficient implementation and analytics mechanisms. Currently, existing approaches mainly address historical spatio-temporal data to deal with the volumetry aspect [14,16,29] or stream time series to deal with continuous queries [3,11,25]. While these systems are efficient for batch or stream time series, there is a lack of a unified approach that combines batch and stream processing and tackles the unique characteristics of spatial mobile sensing data streams modeling and processing. In this paper, we present a prospective data model and a query processing module for sensed data streams. Our approach offers a high-level logical view of **S**patio-**T**emporal **D**ata **S**eries (**STDS**) as well as an internal physical model that combines aggressive indexing and partitioning over time and space to dissolve the heterogeneity and the variety of data. We introduce an incremental query processing approach within a distributed framework to take into account the real-time processing of continuous queries, the large volume, and the high velocity of data. Our contributions are as follows:

- A high level logical view of STDS and a multi-granular physical data model that combines temporal and spatial partitioning.
- An extension of a unified distributed framework for big stored and stream data.
- A query optimizer within an incremental query processing model that offers a set of customized transformations rules for the optimization of spatio-temporal queries.

The rest of this paper is organized as follows. Section 2 discusses the major challenges related to big sensor data. Section 3 presents the related work while Sect. 4 provides an overview of our system architecture. In Sect. 5, we explain the details of the data model. The query processing workflow is presented in Sect. 6, and Sect. 7summarizes the paper and provides some directions for future works.

2 Challenges of STDS Management

Data measured by mobile sensors can be represented by multivariate time series with a focus on the spatial dimension in addition to the temporal one. Such trajectory data denote the paths traced by sensors moving in space over time. Besides measuring the series of geographical positions over time, trajectory data may also contain additional time-dependent variables such as the measurements of surrounding air pollution of the moving object. This large volume of data exhibits a number of challenging characteristics:

Spatial and Temporal Autocorrelation. From the modeling view, a distinctive aspect of such data series is the spatial autocorrelation, meaning that close objects tend to be more similar than distant objects. The same holds for consecutive observations on the same device. As a result, collected data from moving objects cannot be modeled as independent data, and specific algorithms taking into account the correlation between observations occurring in different locations, or between different periods of time need to be considered.

Data Heterogeneity. A notable characteristic is the heterogeneity in space and time. The strength of MCS relies on the usage of different types of sensors designed by different manufacturers that may vary in their sensitivity, sampling frequency, and noise immunity. The data collected from all sensing object should be merged, which could lead to measurements at irregular time intervals and missing data problems. We could observe timestamps that are closely spaced or too sparse in different cases. In fact, some sensors may be offline for hours or stay idle when the device is static (some sensors use the accelerometer to control the sampling rate), they can switch to a burst mode in some situations (increasing the sample frequency more than the normal rate) or stop transmitting the data if the variation is less than a predefined threshold, we could also get different sensor position resolutions. Such heterogeneous data sources should be taken into account in the model, and a harmonized view on the data is highly desirable in order to facilitate their processing and analytics.

Multi-Granularity. Besides, one of the most fundamental characteristics of mobile sensor data is the diversity of their granularity, both under the temporal and spatial dimensions. The temporal domain is typically represented at different time granularities. The spatial entity can be represented using a hierarchical representation that describes the subdivision of the spatial domain into different regions or cells. Combining multiple datasets with several granularities or changing the granularity of a dataset are important analysis tasks that we intend to deal with. Thus, we need to define a multi-granularity framework that takes into account the definition of the spatial and temporal granularities.

Data Volume. Huge amounts of data are being collected continuously from ubiquitous sensor-enhanced mobile devices (as many as the number of equipped holders) in different geographical areas. This requires leveraging big data processing techniques (e.g., Hadoop or Spark) to achieve in-depth understanding, and provide useful information.

Data Velocity. New rows in STDS are typically inserted in recent time intervals as appended rows. Thus, it is necessary to maintain efficient storage structures to handle the velocity of newly arriving data. The commonly used technique in online systems is to consider recent data as more relevant and flush old data. The limitation of such an approach is that some historical data is deleted, as a result, it misses the opportunity to process such data.

Continuous Queries. Due to the continuous processing of sensor data, spatio-temporal queries should be evaluated continuously, which necessitates an incremental processing paradigm. Traditional approaches for processing spatio-temporal data rely on historical data. While analyzing such archived data is important, it lacks the real-time processing of continuous queries. We need a platform that integrates a range of big data technologies to combine the processing of historical and real-time data. A new system architecture that handles massive volume of spatio-temporal data, covers the unique characteristics of sensor data and integrates batch and dynamic processing is necessary.

3 Related Work

Nowadays, sensor data processing can be oriented towards two perspectives: either an offline approach for querying historical data or an online approach for real time queries.

3.1 Offline Processing of STDS

Considerable research efforts have been devoted to offline management and analysis of big trajectory data (multi-dimensional time series) [12,13,27,34]. These works are characterized by a complete storage of large historical data. Such data is used for offline analysis and knowledge discovery. Depending on the application type and the queries, the system tries to optimize query processing over the entire data. The key idea is to use a partitioning mechanism and distribute query processing among multiple nodes using distributed systems, such as Hadoop or Spark. Most of current works are oriented towards exploiting spatial indexes to design efficient methods for optimized query processing while preserving spatial locality. The objective is to tune the system and optimize spatio-temporal queries by making the best use of existing spatial and temporal indexes. There are three approaches for indexing trajectory data. The first approach is to consider the time dimension as the first dimension besides the spatial location. It divides the time dimension into multiple intervals and builds a spatial index (e.g., R-tree) for the trajectories in each time interval [2], or partitions spatial data within each time interval into spatial chunks and loads only relevant chunks for processing [29]. The second approach is to avoid discrimination between the spatial and the temporal dimensions using 3D space-filling curves techniques as proposed in Geomesa [16]. This approach allows to map spatio-temporal points into a single dimension and ensure data locality. It is efficient for queries that combine both temporal and spatial criteria. This category includes also the variations and extensions of R-trees: 3D R-tree, TB-tree, STR-tree [24]. The third approach is to alternate time and space. For example, T-PARINET [26] is based on a combination of spatial partitioning and B+-tree local indexes. Except T-PARINET, the aim of such systems is indexing large historical trajectory data. Therefore, they remain limited when it comes to the support of real time application which central goal is to minimize update costs and to support continuous queries. Moreover, they do not encompass other dimensions than space and time, as required in MCS context.

Another related area of research is distributed time series management systems, a survey on existing systems can be found in [19]. Flint [14] is a time series library for Apache Spark, it proposes interesting features for time series manipulation (temporal join, aggregation, moving average ...). It builds a time series aware data structure (*TimeSeriesRDD*) that allows to associate a time range to each partition and preserves the temporal order of data. Flint does not consider the spatial dimension and the real time nature of queries. It also lacks optimization techniques for temporal queries. However, in real life applications, data collected from sensors is represented by multi-dimensional time series where one dimension corresponds to the spatial location traversed by moving objects over time. TimeScaleDB [29] extends PostgreSQL query planner, data model, and execution engine to support SQL queries on time series data. Internally, it splits tables into chunks, each chunk corresponds to a specific time interval and a region of the partition key's space (using hashing). The created partitions are disjoint, which allows the query planner to select only required chunks to resolve a query. While apt at scaling SQL queries on large volume of data, this system is not designed for real time operations, as it lacks the ability to handle continuous queries and stream processing.

3.2 Online Processing of STDS

There are several commercial solutions for stream processing such as Samza [25], Flink [11], Spark Structured Streaming [3,32], Storm [30]. While these systems generally support high ingestion rate and continuous queries, they are not designed for spatial time series. Academic architectures [1,21,31] were proposed in the literature focusing on streaming data and ignoring historical data. The PLACE [22] server is a data stream management system that supports continuous query processing of spatio-temporal streams. It employs an incremental evaluation paradigm that allows to continuously update the query answer and proposes high-level algorithms for continuous spatio-temporal queries. Zhang et al. [33] extends Apache Storm to process data streams for moving objects, they employ a distributed spatial index to process continuous queries (e.g., continuous kNN). SCUBA [23] allows continuous spatio-temporal queries on moving objects. It proposes clustering techniques to group moving objects and queries into moving clusters based on common spatio-temporal properties to optimize query execution. However, the massive volume of historical trajectory data have exceeded the capacities of such streaming architectures. Management and indexing aspects of large volume of data were not their main concern. Batch processing is still needed for data analysis on historical data. Besides, for some queries, all the data is necessary to ensure a more accurate query result. As a result, a common strategy is to use a hybrid architecture that combines stream and batch processing.

3.3 Unified Approach for STDS Management

The growing volume of spatial data series and the rapid increase of the velocity of sensor data streams accelerate the need for a big data architecture that offers continuous processing of data streams. With the emergence of the volume and the velocity issues, Marz introduced the *Lambda architecture* [21] to handle query processing in a scalable and a fault-tolerant way. It is composed of three layers: the batch layer, serving layer, and speed layer. The batch layer is responsible for processing historical data for batch analysis. The speed layer focuses on analyzing incoming streaming data in near real-time and the serving layer aims at merging the results from the previous two layers. This architecture allows to handle large-scale data and integrate batch and real-time processing within a single framework. While the *Lambda architecture* achieves its goal, it comes with high complexity and redundancy. Kreps [20] discussed the disadvantages of the *Lambda architecture* and presented a new approach for real time data processing named the *Kappa architecture*. This architecture favors simplicity by merging the batch and streaming layers and avoiding data replication. Inspired by the *Kappa architecture*, Spark Structured Streaming [3] is a streaming computation system that combines batch and stream processing using the same code. Based on the *Lambda architecture*, PlanetSense [28] is a generic platform for gathering geospatial intelligence from real time data (e.g., social media, passive and participatory sensing). It combines the power of archived data and the dynamics of real time data for spatio-temporal analytics. However, such an architecture inherits the limits of the *Lambda architecture*, as it needs to implement the transformation logic twice, once in the batch system and once in the stream processing system. In contrast to our proposal, PlanetSense does not propose to manipulate temporal operations specific to time series and lacks a representative data model for time series data taking into account the heterogeneity of data.

4 System Overview

Figure 1 describes our vision of a unified framework for processing batch and streaming STDS.

Data Sources. There are two types of data ingestion in the system. The first is batch data ingestion (finite datasets) that consists of loading a previously acquired data (e.g., loading the data from a previous campaign as a cold start of the data store). The second type corresponds to streaming data (infinite datasets) collected from current campaigns. Notice that we archive the acquired data for further batch analysis, when needed.

Core Engine. The main data processing is performed within the core engine. It contains three components: the query parser, the optimizer and the indexing component (as detailed hereafter). Internally, we rely on existing distributed and streaming engines to generate parallel dataflows. Our target is not a pure streaming framework that requires hard real-time handling. Therefore, we have chosen Spark Structured Streaming as a back-end and exploit the parallel receivers of

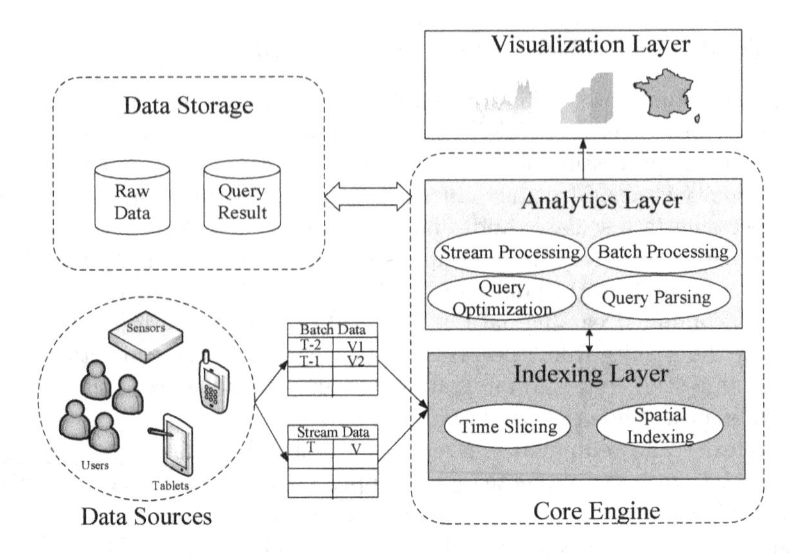

Fig. 1. System architecture

Spark Streaming to collect data from different data sources. Our queries are processed using the Spark's micro-batch model, which processes data streams as a series of small batch jobs.

Query Parser. The query parser is responsible for checking and validating the query syntax. It translates an algebraic expression into a set of transformations on Spark Streaming DataFrame, such as *selection*, *projection*, *temporal join*, *shift*, *aggregation* to convert values to coarser or finer granularity. These operators form an algebra, as an extension of the relational algebra to account for the semantic of STDS (see Sect. 5 for more details). Subsequently, we propose to extend the Spark Streaming DataFrame API to support spatio-temporal operations.

Query Processing. Our query optimizer creates a series of incremental execution plans from a streaming logical plan. The idea is to inject appropriate optimization rules to obtain optimized execution plans. Rule-based optimization in our context exploits spatial indexing and time slicing to access the smallest possible number of partitions, avoiding cartesian product (in case of temporal or spatial joins) or performing selection on required time intervals as early as possible ... Indeed, our query optimizer injects optimization rules to avoid scanning all the data series and eliminates records that do not contribute to the query result. Our query processing module uses the Spark streaming DataFrames/Datasets API to combine batch and stream processing. This data structure allows to represent bounded data as well as streaming data. The advantage is that we can apply the same operations on static and streaming DataFrames.

Indexing. We integrate the concept of indexing to achieve better query performances. We employ time slicing to divide the input data into multiple slices that are distributed on their time granularity. Data within each slice is further divided into sub-slices based on spatial indexing techniques. The key idea is to consider the temporal dimension first, which is important for time series analysis. Details about our physical model are presented in Sect. 5.3.

Data Storage. In order to provide accurate and real-time analysis, we maintain different data storage for raw data and query (continuous) results. Raw data refers to incoming stream time series data persisted for batch analysis using *Parquet* format. New streaming data are maintained in memory until it exceeds a memory threshold (specified by the user). Once the threshold is reached, it is flushed to a separate data storage. This is done periodically using an *append mode* to the DataFrame. Doing so, data are blocked in output files which size is controlled, avoiding the inefficient multiplicity of small files resulting from the default mode.

5 Data Model

As a running example, we consider a database derived from Polluscope[1]. A cohort of volunteers is collecting sensory data each in a different STDS, possibly at different granularities. For instance, GPS data are acquired at a higher frequency than air pollutants such as PM2.5 and NO2.

5.1 Preliminaries

The notion of granularities has been deeply studied in the literature, Bettini et al. [4–6] define the temporal granularity as a partition of the time domain.

Definition 1 *(Temporal Granularity). Formally, a temporal granularity g_T is a function from an ordered set I_T to the power set of the temporal domain T such that:*

$$\forall i, j, k \in I_T, (i < k < j \wedge g_T(i) \neq \emptyset \wedge g_T(j) \neq \emptyset \Longrightarrow g_T(k) \neq \emptyset)$$

$$\forall i, j \in I_T, (i < j \Longrightarrow \forall x \in g_T(i)\ \forall y \in g_T(j)\ x < y)$$

Typical examples of temporal granularities are days, weeks, months. $g_T(i)$ are called temporal granules of the granularity g_T. The first condition states that the subset of the set that maps to non-empty subsets of the time domain is contiguous. The second condition states that granules do not overlap and that their order is the same as their time domain order. Besides, Camossi et al. [10] define the spatial granularity as a mapping from an index set to subsets of the spatial domain (i.e. a set of 2−dimensional points).

Definition 2 *(Spatial Granularity). Formally, a spatial granularity g_S is a function from an ordered set I_S to the power set of the spatial domain S such that:*

$$\forall i, j \in I_S, (i \neq j \wedge g_S(i) \neq \emptyset \wedge g_S(j) \neq \emptyset \Longrightarrow intersects(g_S(i), g_S(j) \neq \emptyset)$$

[1] A French project to build a participative observatory for the surveillance of individual exposure to air pollution and health effects. ANR-15-CE22-0018 Grant. http://polluscope.uvsq.fr.

Typical examples of spatial granularities are pixels of different sizes, or a spatial hierarchy such as administrative subdivisions of a country. $g_S(i)$ are called spatial granules of the granularity g_S.

5.2 Logical Data Model

Definition 3 *(Time Series). We define a time series as an infinite sequence of values where a value is a couple (t, v) where $t \in T$ is a timestamp (at a given granularity) from a time domain T with discrete time units in increasing order and v is a vector (v_1, ..., v_n) where each value is a measurement or scalar value, v is an n-tuple of a fixed size.*
The left side Fig. 2 shows an example of a time series that records the time t and the corresponding values (e.g., PM2.5, PM10, NO2) captured in a vector v.

Definition 4 *(Spatio-Temporal Data Series). We define a Spatio-Temporal Data Series (STDS) as a time series where the location (e.g., latitude and longitude) belongs to the vector v.*

Definition 5 *(Empty Value). The empty value (denoted '!') means that there exist no value. As a time series R is an infinite sequence of vectors, $\forall v \in R, time(v) \in T$. If v is not defined at a given time, then $v = !$.*

Definition 6 *(Unknown Value). The unknown value (denoted '?') means that the value is undefined. It is equivalent to the NULL value in the relational model.*

Using our model, a time series constitutes a linear space vector (a collection of vectors). This mathematical structure allows us to apply basic vector space operations (*plus*, *minus*, *scale*). Thus, time series could be added (denoted +), multiplied (denoted $*$) by numbers (scalar) or even combined linearly in expressions such as $TS1 + s * TS2$ (s being a real). Our model includes also an extension of the relational algebra operators to the support of time series. To this end, we revisit the operators such as temporal *selection, projection, join, union, intersection, aggregation* as follows.

Definition 7 *(Temporal Selection). The temporal selection applied to a time series is a time series where we replace the original value with an empty value (!) if the predicate is not satisfied. Denoting (t, v) the entry t of vector v of the processed time series:*

$$TSel_{pred}(R) = \{(t, v) \mid (t, v) \in R \wedge pred(v)\} \cup \{(t, !) \mid (t, v) \in R \wedge\, !pred(v)\}$$

Example: Selection of air pollutants exceeding a certain threshold (e.g., 50).

$$TSel_{vi>50}(R)$$

Definition 8 *(Window Selection). We define a window selection operator as a transformation that selects values satisfying a temporal predicate w. Formally:*

$$WSel_w(R) = \{(t, v) \mid (t, v) \in R \wedge overlaps(t, w)\}$$

Example: Selection of air pollutants during a specific time period.

$$WSel_{[01/06/2019,15/06/2019]}(R)$$

Definition 9 *(Temporal Projection). We define a temporal projection operator as a transformation that applies a function to each value of the time series it is applied on. Formally:*

$$TProj_f(R) = \{(t, (v_1, ...v_k)) \mid (t, v) \in R \wedge v_i = f(v) \ for \ i \in \{j_1, j_2, ..., j_k\}\}$$

Where f is a linear function that preserves vector addition and scalar multiplication.
Example: Projection and multiplication of PM10 values by 2.

$$TProj_{v2*2}(R)$$

$$TProj_{v2*2}(R)$$

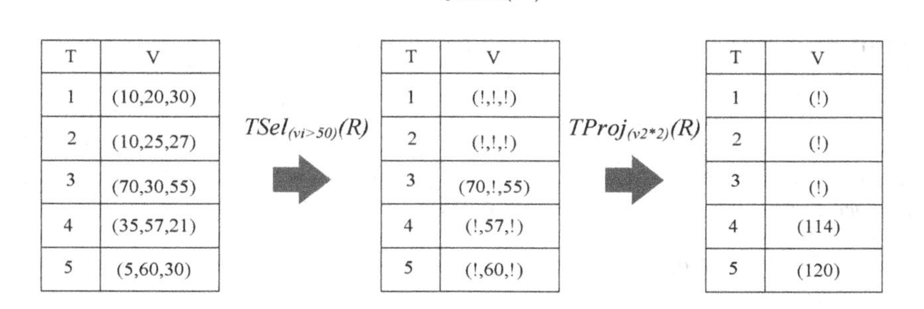

Fig. 2. Selection and Projection with Examples

Definition 10 *(Shift). We define a shift operator as a transformation that applies a shift (future or past) to each timestamp of the time series. Formally:*

$$Shift_\delta(R) = \{(t', v) \mid (t, v) \in R \wedge t' = t + \delta\}$$

Example: Shift the environmental time series by 1 day in order to compare the exposure to pollutants between two consecutive days.

$$Shift_{1day}(R)$$

Definition 11 *(Temporal Intersection, Difference & Union). We define an adaptation of the relational intersection, difference and outer union as follows:*

$$S_1 \cap S_2 = \{(t, v) \mid (t_1, v_1) \in S_1 \wedge (t_2, v_2) \in S_2 \wedge t = intersects(t_1, t_2) \wedge v = v_1 = v_2\}$$

$$S_1 - S_2 = \{(t, v) \mid (t, v) \in S_1 \wedge (t, v) \notin (S_1 \cap S_2)\}$$

$$S_1 \cup S_2 = \{(t, v) \mid (t, v) \in (S_1 - S_2) \vee (t, v) \in (S_2 - S_1) \vee (t, v) \in (S_1 \cap S_2)\}$$

Definition 12 *(Temporal Aggregation). Let R be a time series with a timestamp attribute t, f be an aggregation function (e.g., sum, count, avg) that takes a set of values as an argument and applies an aggregation. We define a set* G_T *that contains all granules* τ *of granularity* g_T *such that* $G_T(R) = \{\tau | \tau \in cast(t, g_T) \wedge t \in R\}$*. Each element of* G_T *defines a partition* $S(\tau, R)$ *of R such that:*

$$S(\tau, R) = \{v \mid (t, v) \in R \wedge overlaps(time(v), \tau)\} \tag{2}$$

The temporal aggregation is defined as:

$$TAgg_{(g_T, f)}(R) = \{(\tau, f(S(\tau, R))) \mid \tau \in G_T(R)\}$$

The set G_T *(e.g.,* $G_T = \{March, April, June, July\}$*) ranges over the granules of a granularity* g_T *(e.g., months).* $S(\tau, R)$ *collects all the values* $v \in R$ *such that* $time(v)$ *is overlapping* τ*. A result tuple is then produced by extending* τ *with the result of the aggregate function f that is computed over each element of* $S(\tau, R)$*. Example: Compute the monthly average of all air pollutants.*

$$TAgg_{(months, avg)}(R)$$

Definition 13 *Window Aggregation. This operator allows to partition and aggregate values over time, based on a moving window w. We define a set* $W_T(R) = \{\tau_1, \tau_2, \tau_3, ..., \tau_q\}$ *that contains the collection of time intervals dividing the time horizon* t_I *of R into sub-intervalls of the size of the window w such that* $t_I = \bigcup_{i=1}^{q} \tau_i$*. Each element of* W_T *define a partition* $S(\tau, R)$ *as defined in Eq. 2. Thus, we define the window aggregation as:*

$$WAgg_{(w, f)}(R) = \{(\tau, f(S(\tau, R))) \mid \tau \in W_T(R)\}$$

Example: Compute the average over a half an hour moving window of a specific air pollutant.

$$WAgg_{(30min, avg)}(R)$$

Definition 14 *(Temporal Join). The temporal join between two time series R and S allows to append each row in R with the row in S at the same time values. Formally:*

$$TJoin(R, S) = \{(t, v) \mid (t_1, v_1) \in R \wedge (t_2, v_2) \in S \wedge v = v1 \oplus v2$$
$$\wedge t = intersects(t1, t2) \wedge t \neq \emptyset\}$$

Example: Match the environmental time series (R) with the GPS data (L).

$$TJoin(R, L)$$

Definition 15 *(Shift Temporal Join). The future (past) temporal join between two time series R and S allows to append each row in R with the closest future row in S at or after (before) a time interval* δ*. It is a redefinition of the temporal join and the shift operators. Formally:*

$$TJoin_\delta(R, S) = \{(t, v) \mid (t_1, v_1) \in R \wedge (t_2, v_2) \in S \wedge v = v1 \oplus v2$$
$$\wedge t = intersects(t1 + \delta, t2) \wedge t \neq \emptyset\}$$

Example: Match the environmental time series (R) with the shifted humidity (H) which variation may impact the values in R after 3mn.

$$TJoin_{3mn}(R, H)$$

Definition 16 *(Spatial Aggregation). We define the spatial counterpart of G_T denoted G_S such that $G_S(R) = \{s | s \in cast(location(t), g_S) \wedge t \in R\}$ where g_S is a spatial granularity and $location(t)$ gives the spatial element of a timestamp $t \in R$. Each element of G_S defines a subset of values as follows:*

$$S_{g_S}(s, R) = \{v \mid (t, v) \in R \wedge overlaps(location(t), s)\}$$

Similar to the temporal aggregation, the spatial aggregation is defined by the following expression:

$$SAgg_{(g_S, f)}(R) = \{(s, f(S_{g_S}(s, R))) \mid s \in G_S(R)\}$$

Example: For each country, compute the highest value of a specific air pollutant.

$$SAgg_{(countries, max)}(R)$$

We could also propose other operators such as a *split* operator which subdivides a single time series into multiple segments, or a *clustering* that partitions a time series according to consecutive similar values, or spatio-temporal join by adding a spatial predicate.

5.3 Physical Data Model

Physically, an infinite sequence of values cannot be stored. As in [18], at storage level, we propose a discrete model of time series data to implement the infinite sequence as in the logical (abstract) model. We physically encode time series data as a set of sequences with specific metadata. As streaming time series enters our framework, each timestamp is represented by a positive integer named Epoch and a granularity. The granularity is chosen based on the original precision of data (e.g., 2019/06 corresponds to a one-month granularity). To reduce the storage overhead, for each sequence of a time series, we only store an array of values omitting the actual timestamp. In fact, we associate a start Epoch value s, and a granularity g as metadata. Then the timestamp of the ith values of the array can be recovered by the simple formula $s + i * g$. If the analysis needs to operate over raw time format, then we use the metadata to calculate the row number of the corresponding values. Our model allows to manage missing data, for example, if there is no value between two sequences, at the storage level nothing is stored, at the logical level, it is represented by an empty value denoted as '!'.

To summarize the first phase of our proposed physical model, time series data is organized as sequences of lists where each list is represented by a granularity. This allows to structure data in a temporal hierarchy where each granularity is represented as a level in the hierarchy. In order to speed-up the access and the filtering on time series, and in particular to STDS data, we propose a model that alternates temporal and spatial indexing. We envision a two-level partitioning scheme, where the first level follows a global index, and the second depends on a second index. At the lowest level, the data will be further divided into even smaller spatial sub-partitions called *buckets*. For instance, the first level can leverage a spatial index while the second follows a temporal partition, and the *buckets* could be based on the order of spatial filling curves. The way the spatial and temporal dimensions will be alternated is not yet decided and may require fine tuning to adapt to the data and the query profile. For query processing, our system will only scan the content of the spatial *bucket* (e.g., `/sid=20`) that can be accessed from the temporal partition (e.g., `/ts.parquet/nump=10`) that contains a range of Epoch indices. This physical organization is inspired by the physical optimization employed for large volume of astronomical data proposed in [7,9]. It has proved its efficiency, as it processes the query in a way that makes it efficient to retrieve the contents of required spatial buckets, obviate scanning irrelevant partitions and allow fast aggregation queries for granularity conversions. We use an *append* save-mode to load additional data while avoiding to overwrite existing data. Data is archived in time-ordered partitions. New incoming time series naturally arrives time-ordered, this allows new data to be appended to existing partitions rather than having to re-sort data into previously stored partitions.

6 Query Processing

Spark Structured Streaming is based on the Spark Catalyst extensible optimizer [3] which allows adding new optimization techniques. Our knowledge of the Spark Catalyst optimizer helped us in designing a new query processing model for STDS. Figure 3 represents our query processing workflow that consists of three major steps: Extended Analysis, Incrementalization and Extended logical-physical optimizations. The input is an algebraic expression that is translated into a set of transformations on Spark Streaming DataFrames by the query parser. This allows to leverage the optimizations offered by the Catalyst optimizer and to inject new optimization techniques.

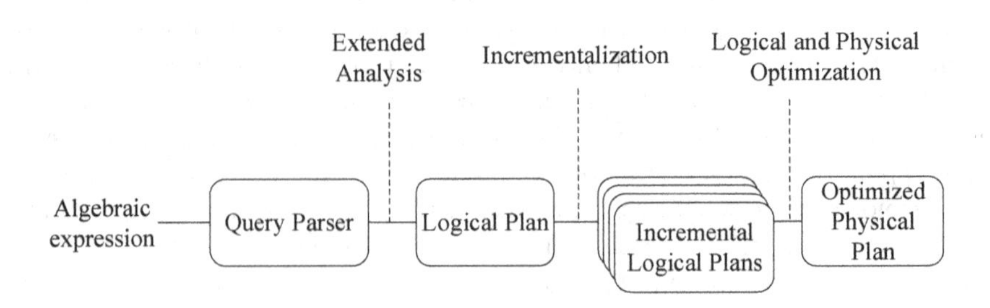

Fig. 3. Query processing workflow

Extended Analysis. The Extended Analysis extends the Spark Structured Streaming analysis to resolve spatio-temporal operations. It validates the query and resolves the attributes and data types. We intend to extend the catalyst optimizer to inject resolution rules in order to transform an impossible-to-solve plan into an analyzed logical plan.

Incrementalization. The next step is to use the Spark Structured Streaming incrementalization technique that allows to continuously update results in response to new data. The main idea is to only report the changes in the query result since the last trigger. This ensures a fast query evaluation because we limit the amount of data to get the query result. We rely on the concept of incremental algorithms to transform queries into trees of traditional logical operators (e.g., join, filter ...).

Extended Logical and Physical Optimizations. The objective of this step is to exploit logical optimization to transform the logical plan into an optimized logical-physical plan using indexing and partitioning. It allows to solve the filtering phase of some proposed queries by either transforming the temporal join query into an equi-join or by filtering the relevant partitions required by spatial or temporal ranges... Our query processing module applies multiple optimization rules to map a query plan to a semantically equivalent plan. Some examples of optimization rules that could be included:

- Temporal Partition Pruning. It consists in determining the temporal partitions that need to be scanned, and hence the epoch indices that can be pruned, to get the query result. Such a rule could be applied for selection queries and allows to access relevant values using their rows number.
- Spatial Index PushDown. This optimization allows to inject new filters in the query plan to eliminate loading spatial objects that do not contribute to the query result. This ensures that such filters are applied at the low level rather than dealing with the entire temporal partition.
- Avoiding Cartesian Product. Temporal join queries can be conceptually formulated as cartesian products. A possible optimization is to replace this product by an equi-join on Epoch indices. The trick is to take inspiration from our spatial join algorithm [8] by shifting all objects of a reference dataset on the fly and transforming the start epoch values. Then, a simple equi-join query on epoch indices becomes sufficient to generate candidate results.

7 Conclusion

In this paper, we presented a unified framework for mobile sensing big stored and stream data. Our framework extends Spark Structured Streaming with the adaptation of data organization and the injection of various optimization rules to optimize processing of stream and historical data series. We also presented a logical data model for STDS and a multi-granular internal data model to take into account the heterogeneity of data. We presented the key query processing workflow of our framework to support incremental algorithms and logical/physical optimizations. Currently, we are working on an integrated prototype within Spark and Catalyst to support an important application branch of MCS which is air quality sensing where air pollution is monitored using multi-sensor devices within the Polluscope project. To deal with the heterogeneity of data, we are currently working on the integration of spatial and temporal disaggregation techniques using ancillary data. These techniques allow a high-resolution and unified output in both the temporal and spatial dimensions.

Acknowledgements. This work benefited from the support of the project POLLUSCOPE ANR-15-CE22-0018 of the French National Research Agency (ANR). It has been also supported by the MASTER project that has received funding from the European Union's Horizon 2020 research and innovation programme under the Marie-Slodowska Curie grant agreement N. 777695.

References

1. Abadi, D.J., et al.: Aurora: a new model and architecture for data stream management. VLDB J. **12**(2), 120–139 (2003)
2. Alarabi, L., Mokbel, M.F., Musleh, M.: St-hadoop: a mapreduce framework for spatio-temporal data. GeoInformatica **22**(4), 785–813 (2018)
3. Armbrust, M., et al.: Structured streaming: a declarative API for real-time applications in apache spark. In: Proceedings of the International Conference on Management of Data (2018)
4. Bettini, C., Jajodia, S., Wang, S.: Time Granularities in Databases, Data Mining, and Temporal Reasoning. Springer, Heidelberg (2000). https://doi.org/10.1007/978-3-662-04228-1
5. Bettini, C., Wang, X.S., Jajodia, S.: A general framework for time granularity and its application to temporal reasoning. Ann. Math. Artif. Intell. **22**(1–2), 29–58 (1998)
6. Bettini, C., Wang, X.S., Jajodia, S., Lin, J.L.: Discovering frequent event patterns with multiple granularities in time sequences. IEEE Trans. Knowl. Data Eng. **10**(2), 222–237 (1998)
7. Brahem, M., Yeh, L., Zeitouni, K.: Efficient astronomical query processing using spark. In: Proceedings of the 26th ACM SIGSPATIAL International Conference on Advances in Geographic Information Systems (2018)
8. Brahem, M., Zeitouni, K., Yeh, L.: Hx-match: In-memory cross-matching algorithm for astronomical big data. In: International Symposium on Spatial and Temporal Databases (2017)
9. Brahem, M., Zeitouni, K., Yeh, L.: Astroide: a unified astronomical big data processing engine over spark. IEEE Trans. Big Data (2018)
10. Camossi, E., Bertolotto, M., Bertino, E.: A multigranular object-oriented framework supporting spatio-temporal granularity conversions. Int. J. Geogr. Inf. Sci. **20**(05), 511–534 (2006)
11. Carbone, P., Katsifodimos, A., Ewen, S., Markl, V., Haridi, S., Tzoumas, K.: Apache flink: Stream and batch processing in a single engine. Bull. IEEE Comput. Soc. Techn. Committee Data Eng. **36**(4), (2015)
12. Ding, X., Chen, L., Gao, Y., Jensen, C.S., Bao, H.: Ultraman: a unified platform for big trajectory data management and analytics. Proc. VLDB Endowment **11**(7), 787–799 (2018)

13. Fang, Y., Cheng, R., Tang, W., Maniu, S., Yang, X.: Scalable algorithms for nearest-neighbor joins on big trajectory data. IEEE Trans. Knowl. Data Eng. **28**(3), 785–800 (2015)
14. Flint. https://github.com/twosigma/flint. Accessed May 2019
15. Ganti, R.K., Ye, F., Lei, H.: Mobile crowdsensing: current state and future challenges. IEEE Commun. Mag. **49**(11), 32–39 (2011)
16. Geomesa. https://www.geomesa.org/. Accessed May 2019
17. Guo, B., et al.: Mobile crowd sensing and computing: the review of an emerging human-powered sensing paradigm. ACM Comput. Surv. (CSUR) **48**(1), 7 (2015)
18. Güting, R.H., Behr, T., Düntgen, C., et al.: Secondo: a platform for moving objects database research and for publishing and integrating research implementations. IEEE Data Eng. Bull. **33**(2), 56–63 (2010)
19. Jensen, S.K., Pedersen, T.B., Thomsen, C.: Time series management systems: a survey. IEEE Trans. Knowl. Data Eng. **29**(11), 2581–2600 (2017)
20. Kreps, J.: Questioning the lambda architecture. Online article, July (2014)
21. Marz, N., Warren, J.: Big Data: Principles and Best Practices of Scalable Real-Time Data Systems. Manning Publications Co., New York (2015)
22. Mokbel, M.F., Xiong, X., Aref, W.G., Hambrusch, S.E., Prabhakar, S., Hammad, M.A.: Place: a query processor for handling real-time spatio-temporal data streams. In: Proceedings of the International Conference on Very Large Data Bases-Volume 30 (2004)
23. Nehme, R.V., Rundensteiner, E.A.: SCUBA: scalable cluster-based algorithm for evaluating continuous spatio-temporal queries on moving objects. In: Ioannidis, Y., et al. (eds.) EDBT 2006. LNCS, vol. 3896, pp. 1001–1019. Springer, Heidelberg (2006). https://doi.org/10.1007/11687238_58
24. Pfoser, D., Jensen, C.S., Theodoridis, Y., et al.: Novel approaches to the indexing of moving object trajectories. In: VLDB (2000)
25. Samza. http://samza.apache.org/. Accessed May 2019
26. Sandu Popa, I., Zeitouni, K., Oria, V., Barth, D., Vial, S.: Indexing in-network trajectory flows. VLDB J. **20**(5), 643–669 (2011)
27. Shang, Z., Li, G., Bao, Z.: Dita: distributed in-memory trajectory analytics. In: Proceedings of the 2018 International Conference on Management of Data (2018)
28. Thakur, G.S., Bhaduri, B.L., Piburn, J.O., Sims, K.M., Stewart, R.N., Urban, M.L.: Planetsense: a real-time streaming and spatio-temporal analytics platform for gathering geo-spatial intelligence from open source data. In: Proceedings of the 23rd SIGSPATIAL International Conference on Advances in Geographic Information Systems (2015)
29. TimeScale. https://www.timescale.com/. Accessed May 2019
30. Toshniwal, A., et al.: Storm@ twitter. In: SIGMOD International Conference on Management of Data (2014)
31. Tran, D.A., Hua, K.A., Do, T.T., et al.: A peer-to-peer architecture for media streaming. IEEE J. Sel. Areas Commun. **22**(1), 121–133 (2004)
32. Zaharia, M., Das, T., Li, H., Hunter, T., Shenker, S., Stoica, I.: Discretized streams: fault-tolerant streaming computation at scale. In: Proceedings of the Twenty-Fourth ACM Symposium on Operating Systems Principles (2013)
33. Zhang, F., et al.: Real-time spatial queries for moving objects using storm topology. ISPRS Int. J. Geo-Inf. **5**(10), 178 (2016)
34. Zhang, Z., Jin, C., Mao, J., Yang, X., Zhou, A.: Trajspark: a scalable and efficient in-memory management system for big trajectory data. In: Asia-Pacific Web (APWeb) and Web-Age Information Management (WAIM) Joint Conference on Web and Big Data (2017)

BY

Predicting Fishing Effort and Catch Using Semantic Trajectories and Machine Learning

Pedram Adibi[1(✉)], Fabio Pranovi[2], Alessandra Raffaetà[2], Elisabetta Russo[2], Claudio Silvestri[2], Marta Simeoni[2], Amilcar Soares[1], and Stan Matwin[1,3]

[1] Institute for Big Data Analytics, Dalhousie University, Halifax, Canada
{pedram.adibi,amilcar.soares,stan}@dal.ca
[2] Dipartimento di Scienze Ambientali, Informatica e Statistica, Università Ca' Foscari Venezia, Venezia, Italy
{pranovi,raffaeta,russo,silvestri,simeoni}@unive.it
[3] Polish Academy of Sciences, Warsaw, Poland

Abstract. In this paper we explore a unique, high-value spatio-temporal dataset that results from the fusion of three data sources: trajectories from fishing vessels (obtained from terrestrial Automatic Identification System, or AIS, data feed), the corresponding fish catch reports (i.e., the quantity and type of fish caught), and relevant environmental data. The result of that fusion is a set of semantic trajectories describing the fishing activities in Northern Adriatic Sea over two years. We present early results from an exploratory analysis of these semantic trajectories, as well as from initial predictive modeling using Machine Learning. Our goal is to predict the Catch Per Unit Effort (CPUE), an indicator of the fishing resources exploitation useful for fisheries management. Our predictive results are preliminary in both the temporal data horizon that we are able to explore and in the limited set of learning techniques that are employed on this task. We discuss several approaches that we plan to apply in the near future to learn from such data, evidence, and knowledge that will be useful for fisheries management. It is likely that other centers of intense fishing activities are in possession of similar data and could use the methods similar to the ones proposed here in their local context.

Keywords: Spatio-temporal data · Fisheries · Machine Learning · Semantic trajectories · AIS

The authors would like to thank NSERC (Natural Sciences and Engineering Research Council of Canada) for financial support. This work was partially supported by project MASTER (Marie Sklowdoska-Curie agreement N. 777695), which has received funding from the EU Horizon 2020 Programme.

K. Tserpes et al. (Eds.): MASTER 2019, LNAI 11889, pp. 83–99, 2020.
https://doi.org/10.1007/978-3-030-38081-6_7

1 Introduction

In this paper, we present early results from an ongoing international research project in which mobility data researchers and fishery ecologists collaborate closely. In our project, we explore a unique, high-value dataset that results from the fusion of three data sources: trajectories from fishing vessels, the corresponding fish catch reports (i.e., the quantity and type of fish caught), and relevant environmental data. The goal of this project is to predict the future Catch Per Unit Effort (CPUE) from the past data. CPUE is an indicator of fishing resources exploitation that allows for assessing the pressure of these activities at the ecosystem level. Intuitively, a decrease of CPUE indicates a situation of over-exploitation, a steady CPUE value points out a sustainable exploitation of the fishery resources and an increase of its value corresponds to a healthy and growing population. CPUE is therefore a key indicator for fisheries management since it could help to define the sustainability of the fishing activities in the area of interest: an accurate forecast of CPUE could help decision makers to obtain a sustainable fishing business by adapting the fisheries management plans on the basis of the forecast results. Here we discuss and present early results from the use of Machine Learning techniques to predict the CPUE in the Northern Adriatic Sea.

We believe this research demonstrates the opportunities provided by mobility data analysis to gain insights and evidence that can guide fisheries management decisions. Such decisions will have significant environmental and economic consequences at the regional, national, and eventually global level. Our results are preliminary, both in the temporal data horizon that we are able to explore, and in the broader set of techniques that could be employed on this task. It is likely that other centers of intense fishing activities are in possession of similar data and could use the methods similar to the ones proposed here in their local context.

The Northern Adriatic Sea area, on which our work is based, needs tools and models that can assist fisheries management at the macro scale. This area, known for its very high productivity, is recognized to be one of the most exploited area of the Mediterranean Sea, causing an over-exploitation of the fish resources. In this context, the development of effective fishery management plans is needed to make fishing activities sustainable and ensure a productive and healthy ecosystem. Currently, different management measures are used in the Northern Adriatic Sea (e.g., the permanent ban of trawling activities within 3 nm of the coast, the seasonal biological rest period for trawlers). In this context, forecasting the fishing activities and their catches in space and time represents a step forward to assess the efficiency of these measures and develop new ones.

Recently, several works report the use of mobility-tracking technologies, such as Automatic Identification System (AIS) to monitor fishing activities. In its inception, AIS was primarily designed as a navigational aid to avoid vessel collisions, but nowadays it has become - often due to its open nature - the primary source of data about fishing-related activities. In our setting, we have access to terrestrial AIS data, i.e., AIS data sent by ships and received by ground stations on the Italian coast of Northern Adriatic. Vessels transmit their position at a variable rate, from 2 s up to two minutes. We use AIS data to reconstruct, in time and space, the fishing trajectories. The latter have been enriched with the available environmental data, such as daily surface temperature, chlorophyll-a, and wave height. A further valuable piece of information available for this work is the landing reports of the Chioggia's fishing market, which is the primary market of the Northern Adriatic basin.

The two main research questions that guide this work are: (i) How can we improve our knowledge of the spatio-temporal aspects of the fishing activities in the Northern Adriatic Sea?; and (ii) How can we predict the CPUE for next year?

Data fusion, management, and Machine Learning techniques from the mobility data analysis are applicable to provide evidence-based answers to these questions. In this paper, we focus on the use of semantic trajectories of fishing vessels and predictive modeling using spatio-temporal Machine Learning techniques, and we address mainly the last of the above questions.

The contributions of this work are the proposal of a framework that: (i) integrates the heterogeneous data sources; (ii) extracts knowledge from the integrated data using semantic trajectory modeling; and (iii) applies Machine Learning (e.g., Random Forest) to learn from those semantic trajectories a model for forecasting the CPUE.

This paper is structured as follows: Sect. 2 describes the related work in the literature concerning semantic trajectories and fishing activities forecast. Section 3 illustrates the architecture of the developed system and describes in details the data sources, how the raw data have been fused and incorporated in a semantic model, and the model developed for prediction analysis. Section 4 reports the predictive model results and Sect. 5 draws some concluding remarks and illustrates possible future developments.

2 Related Works

In this section, we discuss related works regarding (i) sea data fusion and semantic trajectories, which is the concept used for enriching complex objects like fishing ships with relevant information; and (ii) fishing activities forecast, which is the final goal of our predictive model.

2.1 Data Fusion of Sea Data and Semantic Trajectories

Combining the AIS, trading transactions, and environmental variables from the vessels into a single representation is challenging. Several strategies were proposed to deal with the fusion of heterogeneous ocean data properly. For example, the paper [18] shows a platform in the maritime vessel traffic domain for discovering real-time traffic alerts by querying and reasoning across numerous streams (e.g., AIS, weather, ice, etc.). The authors use semantic web technologies to integrate heterogeneous data sources. In [6], the authors propose a model for integration and analysis of data for vessel movement in a real-time maritime situation awareness system, also using semantic web techniques and tools. Unlike the previous methods, we model our trajectory data with a semantic model. By considering data sources such as AIS, environmental variables, and landing reports, the trajectory of every fishing vessel becomes a complex object with several data dimensions that are contextual to the movement.

Several semantic models for trajectory data have been proposed, such as the *stops* and *moves* [17], CONSTANT [2], and recently MASTER [13]. This last model is more flexible and expressive since it allows for enriching trajectories with complex objects and it provides not only a conceptual model but also a logical schema in the Resource Description Framework (RDF) and a triplestore based on NoSQL databases for maintaining RDF data.

By following the MASTER semantic model, the trajectory of fishing vessels can be represented as a *multiple aspect trajectory.* The AIS data constitute the sequence of spatio-temporal points. Moreover, the MASTER model introduces the concept of *aspect* which consists of "a real-world fact that is relevant for the trajectory data analysis" [13]. It distinguishes between *volatile* aspects—which are usually associated with the trajectory points, since they vary during the object movement—and *long term* aspects—which do not change during an entire trajectory, and hence they are related to the whole trajectory. For instance, for vessel trajectories, the speed is a volatile aspect, whereas the fishing gear type is a long term aspect.

2.2 Fishing Activities Forecast

The literature on fishing activities forecast is vast and can be analyzed in several ways. From the fishing management perspective, works like [16] propose a seasonal forecast system that combines environmental and fish habitat data (collected by fish tagging) to predict tuna distribution. In [15], the authors, integrate satellite data and statistical models output to investigate the relationship between sea surface temperature and chlorophyll-a concentration, and also define simple methods to forecast potential fishing grounds. The work of [9] tries to forecast 1-month catches considering only the anchovy catches in previous months as inputs. Similarly, to [15,16], we use environmental data (e.g., chlorophyll-a and sea surface temperature). Also similarly to [9] we use fish catch information to predict future catches. Unlike all of them, we use wave height as an environmental variable in our model.

From a perspective that considers the geolocation technology used to track ships, some works use Vessel Monitoring System (VMS) [12,19], satellite images [15] or AIS [10,20,22]. Most of these works focus on training models to forecast when a vessel is performing a fishing activity or not. Different types of fishing ships (e.g., long-liners, purse-seiners, etc.) have different types of movement patterns. Predicting these patterns depends on the training data given to the machine learning model [19,20], or the domain specialist ability to create rules that reflect these patterns [14]. In this work, we use domain knowledge from specialists to determine the activity of vessels (e.g., fishing or not) on their

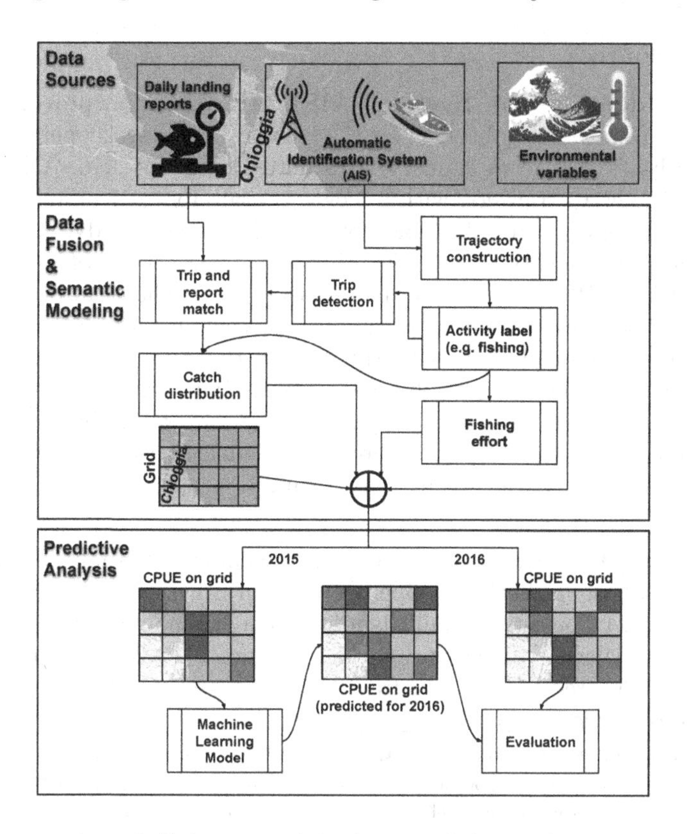

Fig. 1. An overview of all the steps of the framework for predicting fishing catches.

trajectory segments. Based on the knowledge of ranges of fishing speed for different types of fishing gears (e.g., trawlers, long-liners, etc.), we encode the specific rules to detect vessel activities. By exploiting this information, we can compute in a very accurate way the area swept by vessels while fishing, thus allowing for a more realistic estimate of fishing effort and CPUE. To the best of our knowledge, no work in the literature uses a combination of AIS, fishing catch reports, and environmental variables to forecast CPUE.

3 A Framework for Predicting CPUE

In this section, we present the bird's eyes view of the architecture of the system we have developed. We then discuss the individual data sources—AIS data, catch data (landing reports), environmental data—(Sect. 3.1), and the spatio-temporal mapping of data as well as the fusion of the individual sources (Sect. 3.2). The section is rounded up with a brief discussion of the Machine learning method we have selected to build the prediction model (Sect. 3.3). Schematics of the system is shown in Fig. 1.

3.1 Data Sources

Automatic Identification System (AIS). AIS raw data, provided by the Italian Coast Guard, were obtained for the trawl fishing vessels operating in the Northern Adriatic Sea from January 2015 until December 2016. A total of 70 (2015) and 77 (2016) trawlers, with a length overall above 15 m and belonging to the Chioggia navy, were taken into consideration in this study: in particular, small and large bottom otter trawl (SOTB and LOTB), Rapido, one specific kind of beam trawl (RAP), and midwater pair trawl (PTM). The identification of the vessels was performed by matching the data present in the AIS (MMSI code, vessel name and the call sign) with the ones of the European Fleet Register, which supplies specific information on the vessels (i.e., primary and secondary gear, length overall, gross tonnage, etc.). All the data given by the AIS (i.e., data position, speed, time, MMSI) were used to identify the fishing tracks and analyze the fishing activities (fishing, not fishing).

Daily Landing Reports. Landing dataset was obtained from the Chioggia's Fish Market, whose harbor hosts one of the main fishery fleets of the Adriatic Sea. This dataset consists of daily landings (catch amounts in kilogram) for 104 commercial species caught during the biennium 2015–2016 in the Northern Adriatic Sea. The records pertain to 82 fishing vessels, and a total of 17921 fishing trips over the two years.

A graph of total monthly landings for the two years with the contribution of the five most harvested species is shown in Fig. 2. Seasonality of the data is evident by visual comparison of the annual trends. The graph shows zero landing in August for both years, which reflects the fishing ban. It is also visible from the graph that 2015 landing amounts were higher than 2016 for almost all the months.

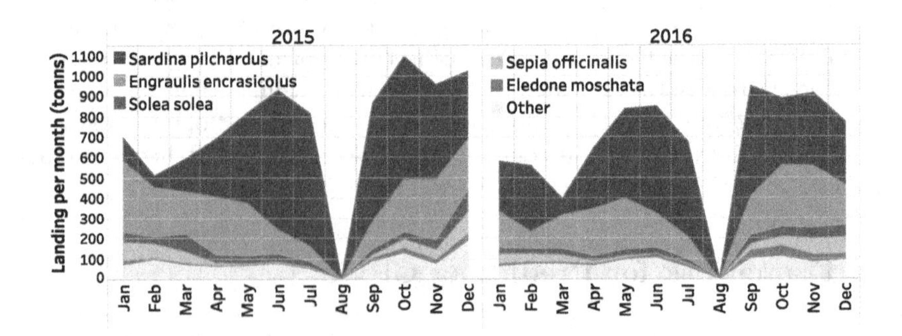

Fig. 2. Chioggia's total monthly landing and species contribution in 2015–16

Environmental Data. We considered the following environmental information in order to enrich the trajectories of the fishing vessels:

- Sea Surface Temperature (in kelvin) [7]
- Sea Daily Chlorophyll-a Concentration (in mg/m^3) [7]
- Spectral significant wave height (in meters) [7]

The sea surface temperature and the chlorophyll-a influence the species distribution, while the wave height affects the fishermen behavior. Hence, adding such semantic information could be relevant for a more accurate prediction of the CPUE indicator. Moreover, the utilization of the sea surface temperature can be helpful to evaluate the effect of climate changes on fishing activities, a hot topic to be considered.

3.2 Data Fusion and Semantic Modeling

Activity Labeling and Trip Detection from AIS Data. Trajectories have been reconstructed by linear interpolation of the raw AIS data. While performing the reconstruction, all implausible points have been discarded. In particular, all the movements that were not physically feasible concerning a maximum possible boat speed were removed.

A trajectory is therefore defined as a sequence of segments, and it is enriched with the following data: MMSI (boat identifier), departure time of the trip (exit from harbour area), departure port, position of the segment with respect to the ports areas, average speed, activity of the boat within the segment and fishing gear (this can be obtained through the MMSI).

The activity attribute is an integer value distinguishing among the following situations: (0): *in port*; (1): *exiting from*; (2): *entering to port*; (3): *fishing*; (4): *navigation*. The *in port*, *exiting from* port and *entering to* port situations can be deduced from the position of the extremes of the segment w.r.t. the port area. In case none of the previous situations occurs, the fishing or navigation activities are established on the basis of the average speed of the boat. More precisely, if the average speed is in the range of the fishing speed of the equipped gear, the boat is assumed to be in a *fishing* phase; otherwise, it is assumed to be in a *navigation* phase.

This trajectory can be modeled as a *multiple aspect trajectory*, following MASTER model [13]. Indeed, as minimum granularity to attach semantic information, we do not consider a single spatio-temporal point, but we annotate segments since we want to highlight the presence of homogeneous trajectory portions, which are the appropriate granularity level for our analyses. Hence the pieces of information we listed above can be classified as

- *long-term aspects*: MMSI, departure time of the trip, departure port and the gear used for fishing because they do not change during the entire trajectory;
- *volatile aspects*: average speed and activity of the boat since they frequently vary during the object movement and they can be associated with a segment.

By using the MASTER model we are able to represent different aspects of our trajectories in a uniform and simple way. Moreover, this representation allows us to perform complex queries merging together spatial, temporal and semantic features. In the rest of the paper, we denote with T the resulting set of multiple aspect trajectories.

The Spatial Grid. Some of the concepts described later in this section (i.e., fishing effort and CPUE) are defined over an area. In order to calculate those values, the area under study is partitioned into a square grid with 5×5 km cell size. Fishing effort and CPUE are then calculated over the individual grid cells. In addition, the grid is used in the calculation of weighted catch distribution (described later), and as the data format for prediction modeling (Sect. 3.3).

Calculation of Fishing Effort over Grid Cells. After reconstructing the trajectories, we proceed with the computation of the *fishing effort*, an essential indicator for monitoring the fishing pressure on an area of interest over time. As mentioned above, we partition the Northern Adriatic Sea into a regular grid. The fishing effort for a cell during a fixed period of time is defined as the ratio between the area of the cell "swept" by vessels while fishing during the given time period and the total area of the cell itself. The swept area depends on the employed gear which can be recovered from a specific dataset where each vessel, identified by its MMSI, is associated with its gear.

In the following we will denote by c a generic cell in the area of interest, by p a time period (could be day, month, etc.) and by g a gear (small and large bottom otter trawl, Rapido and midwater pair trawl).

Definition 1. Let c be a cell, p a time period and g a gear. The *fishing effort* wrt the gear g in the cell c during the time period p is defined as follows:

$$fe(c,p,g) = \frac{(\Sigma_{tr \in T, gear(tr)=g} len(tr,c,p)) * gear\_width(g)}{area(c)} \tag{1}$$

where

- T is the set of multiple aspect trajectories;
- $len(tr,c,p)$ returns the sum of the lengths of the fishing segments of trajectory tr falling in cell c during time period p;
- $gear\_width(g)$ is the width of the net of gear g;
- $area(c)$ is the total area of the cell c.

It is worth noting that we can obtain the total fishing effort in a cell c during a time period p by summing up the fishing effort for each gear. Indeed, for our analyses in Sect. 4 we will use the fishing effort for a particular gear, i.e., Rapido.

Thanks to the reconstruction and the semantic enrichment of trajectories we can compute the lengths of the fishing segments falling in each cell. This allows a more accurate and realistic estimate of the swept area and therefore of the fishing effort.

Assigning Landing Reports to Trips. The landing dataset provided by the Chioggia's fish market contains information about each trading transaction, including the landing date, MMSI of the seller, the species, and the quantity of fish. We have to associate each fish market transaction with a trajectory of the vessel having the specified MMSI. To accomplish this task, for each transaction, we select the vessel trip with the most recent arrival in the port (before 4 PM of the landing date). Such a trip has to respect some constraints: it has to last at least 1 h, and have a minimum length of 2 km, from which, at least 100 m have to be classified as fishing. Arrivals after 4 PM are associated with transactions occurring the next day. Assignment of the quantity (weight) of the fish to a vessel is called a *catch*.

Catches Distribution over Trips. The association of fish catches with trajectories allows us to add a further *volatile* aspect to our multiple aspect trajectories. In fact, we can distribute the fish associated with a trajectory along with its fishing segments. In particular, we can employ two different techniques:

- *uniform* distribution, or
- *weighted* distribution.

In the first case, for each trading transaction, the amount of fish is *uniformly* distributed along with the fishing segments of the corresponding trip. Each fishing segment of the trajectory is associated with a portion of the total amount of fish, proportional to its length.

Of course, the assumption of uniform catch distribution is a simplification of reality. As an improvement, a *weighted* distribution of catches is also considered. The idea behind this approach is that the areas where more vessels are fishing, during a given time period, are more likely to have higher catch rates. A preliminary method based on this idea is implemented as follows. First, the number of distinct vessels that were detected fishing in each grid cell in a time period is computed. Then the amounts of catch over each segment, derived using the uniform distribution, is weighted by the vessel counts in the cell containing the segment. The weights are normalized by the sum of vessel counts in cells that cover all the fishing segments in a trip, so they add up to 1.

Based on this piece of information, we can compute the quantity of fish caught in each cell during a period of time by boats having a particular gear g.

Definition 2. Let c be a cell, p a time period and g a gear, the fish *catch* wrt to the gear g in cell c during the time period p is defined as follows:

$$catch(c, p, g) = \Sigma_{tr \in T, gear(tr)=g}\, quantity(tr, c, p) \tag{2}$$

where

- T is the set of multiple aspect trajectories;
- $quantity(tr, c, p)$ returns the sum of the fish quantities in kilograms associated with the fishing segments of trajectory tr falling in cell c during period p.

Catch Per Unit Effort (CPUE). Catch per unit effort (CPUE) is an indicator of the species abundance in the assessment of fishery resources. This index represents a valid method to evaluate the population trends where, a decrease of CPUE indicates a situation of over-exploitation, a steady CPUE value points out sustainable exploitation of the fishery resources, and an increase of its value corresponds to a healthy and growing population.

Definition 3. Let c be a cell, p a time period and g a gear, the *catch-per-unit-effort (CPUE)* wrt to the gear g in cell c during the time period p is defined as follows:

$$cpue(c, p, g) = \frac{catch(c, p, g)}{fe(c, p, g)} \quad (3)$$

CPUE is, therefore, a key indicator for fisheries management since it gives information on the sustainability of the fishing activities in the area of interest. As a consequence, an accurate forecast of CPUE could help decision makers to maintain a sustainable fishing business by adapting the fisheries management plans based on its forecasted values.

3.3 Predictive Modeling

The objective of the modeling procedure described in this section is the prediction of average *monthly* CPUE values for individual grid cells. First, we describe the modeling data, which consists of *daily* values for the model attributes (or variables) per grid cell. Then, we follow with a brief background on the chosen machine learning method—Random Forest (RF). Finally, we explain how we adjust the temporal granularity to obtain the desired *monthly* output from the model, which is built using the *daily* model attributes.

Modeling Data. Modeling data maps the data onto a spatio-temporal grid, producing *records* (or *instances*) each of which corresponds to a *date*, and a *spatial grid cell* (grid described in Sect. 3.2). Each *record* is comprised of the response attribute—CPUE—and a set of predictive attributes, all pertaining to the same date and grid cell. The predictive attributes are as follows.

- Environmental attributes (described in Sect. 3.1): daily chlorophyll-a concentration, daily sea surface temperature, and daily spectral wave height; each attribute re-sampled over the grid cells.
- Temporal attributes that preserve seasonality: month of year (1–12), day of year (1–365), week of year (1–53), seasons (four quarters starting in January).
- Spatial attributes: latitude and longitude of the grid cell centres.

CPUE is calculated using fishing effort, which in turn, depends on the type of fishing gear. Among the four fishing gears described in Sect. 3.1, Rapido has the largest share in the dataset with 48% of the records. For this reason, we have limited our presentation in this paper to the model trained and tested on data for the vessels with the Rapido gear. Also, in this study, CPUE is calculated based on the total catch amounts for all species in the landing dataset.

As described in Sect. 3.2, CPUE is defined on grid cells with some fishing activity in a given period. Consequently, for a given day, the dataset only includes the fished grid cells. It follows that the size of the modeling dataset is obtained by $\sum_d |C|_d$ where $|C|_d$ is the number of fished cells on date d. Size of the dataset for Rapido vessels amounts to 51262 records over the two year period.

Machine Learning Method. The task of prediction modeling of CPUE presents a regression problem. RF [5] was chosen as the regression method due to the following considerations. It does not require any assumption about the distribution of the model attributes. It can take numeric and categorical attributes, and it does not require scaling of the attributes. Moreover, RF does not result in instability of output values when presented with predictive attributes with values outside the range of the training data. Besides RFs, we have experimented with several other regression methods, e.g. linear regression with LASSO and with the Support Vector Machines. RF, however, outperformed the other methods.

RF is an ensemble learning method based on decision trees, introduced by Breiman. Ensemble leaning methods can improve accuracy, and reduce bias and variance by combining outputs of many base learners [8]. In the case of regression RF, numerous regression decision trees are trained, and the model output is obtained by averaging the outputs of the individual trees. RF uses bootstrap aggregating (Bagging) to construct individual trees that are trained independent of each other. Bagging is an ensemble learning method that trains its base learners on bootstrap samples—samples that are randomly drawn with replacement from a dataset of same size [4]. RF introduces further randomization in the construction of the trees by taking a random subset of the predictive attributes at each node, and selecting one from the random subset to split on.

Adjusting Temporal Granularity. Even though the prediction of average *monthly* CPUE is the main interest of this study, the predictive environmental attributes (e.g. wave height and water surface temperature) affect the fishing activity on a *daily* basis. Therefore, aggregating such attributes on a monthly basis prior to modeling would result in losing the information pertaining to the daily cause and effects of those attributes. To preserve that information, first the RF regression method is performed using the *daily* training data consisting of the 2015 dataset, which produces *daily* predictions for individual grid cells for the year 2016. Then, *monthly* predictions for each cell is calculated by averaging the *daily* predictions for the respective cell over the month. Averaging is used to aggregate the predicted CPUE values, as opposed to summation, because predicted CPUE values are ratios and do not produce a meaningful sum.

The resulting average monthly predictions for individual cells are considered to be the model output. Evaluation of the model is then done against similarly calculated *monthly* CPUE averages per grid cell, using the actual 2016 data. Equation (4) shows the calculation of actual and predicted average CPUE for a given cell c over a given period p (e.g. month)[1], respectively denoted by $y_{c,p}$ and $\hat{y}_{c,p}$,

$$y_{c,p} = \frac{1}{|D_{c,p}|} \sum_{d \in D_{c,p}} cpue(c,d) \quad \text{and} \quad \hat{y}_{c,p} = \frac{1}{|D_{c,p}|} \sum_{d \in D_{c,p}} \widehat{cpue}(c,d) \tag{4}$$

where $D_{c,p}$ indicates the set of all days d in period p for which cell c has a CPUE value. $cpue(c,d)$ and $\widehat{cpue}(c,d)$ are respectively the actual and predicted daily CPUE values for cell c on day d.

4 Experiments and Results

Two models are built and evaluated in this experiment: one for CPUE values calculated using the *uniform* catch distribution, and one for the *weighted* distribution. The distinction between the two distributions is described in Sect. 3.2. The models are trained on year 2015 (training data), then prediction and evaluation are done for year 2016 (test data) of the two year dataset. The reason for splitting the data by year for training and testing is the highly seasonal nature of the data set, as described in Sect. 3.1.

Baseline Model. The baseline, which is used as the benchmark to compare with our RF model, is to simply use the last observed value as the forecast (prediction)—known as naïve forecasting. This baseline is chosen rather than results from other regression models, because it is a standard practice in forecast modeling [11], and it provides a consistent baseline that is independent of the choice of the regression model. In particular, for this experiment, the baseline average monthly prediction of CPUE for a given cell in 2016 is the respective average monthly CPUE for that cell from 2015. Equation (5) illustrates the baseline prediction

$$\hat{y}^*_{c,p} = y_{c,p\downarrow 1} \tag{5}$$

where $y_{c,p\downarrow 1}$ is the actual value for cell c at the same period moved one year backward. For instance, if $p = June2016$ then $p \downarrow 1 = June2015$.

Evaluation Metrics. The metrics used for evaluation are as follows.

- Mean Absolute Error (MAE) is calculated for each period p (i.e. month) as the mean of absolute errors of the predicted average CPUE for all cells in that period. MAE for period p is shown in Eq. (6)

$$MAE_p = \frac{1}{|C_p|} \sum_{c \in C_p} |\hat{y}_{c,p} - y_{c,p}| \tag{6}$$

[1] As pointed out in Sect. 3.3 we consider CPUE only for the gear Rapido. Hence, instead of writing $cpue(c, p, Rapido)$, we simply use $cpue(c, p)$, omitting the gear name.

where C_p denotes the set of all cells with a CPUE value in period p; and $\hat{y}_{c,p}$ and $y_{c,p}$ are respectively predicted and actual CPUE values for cell c and period p. MAE for the baseline prediction is calculated similarly and it is denoted by MAE^*.

- Normalized Mean Absolute Error (nMAE) is calculated for each period as the MAE for that period divided by the mean of the actual CPUE for that period. nMAE for period p is shown in Eq. (7).

$$nMAE_p = \frac{MAE_p}{\mu_p}; \qquad \mu_p = \frac{1}{|C_p|} \sum_{c \in C_p} y_{c,p} \tag{7}$$

- Relative Absolute Error (RAE) is a measure of model performance relative to the baseline model; it is calculated as the ratio of model MAE to the baseline MAE* for a given period [1], as shown in Eq. (8).

$$RAE_p = \frac{MAE_p}{MAE_p^*} \tag{8}$$

Model evaluation metrics for uniform and weighted catch distributions are shown in Table 1. Since interpreting MAE requires information about the magnitude of CPUE, mean of CPUE for each period is also included. Due to the variation of CPUE means, MAEs for different periods cannot be directly compared. nMAE allows for direct comparison of the model performance for different periods, since it is normalized by the period mean. Similarly, RAE allows for comparison of model performance against the baseline for different periods. If RAE equals 1, the model is performing as well as the baseline for that period. RAE of less than 1 indicates that the model performs better than the baseline, and vice versa.

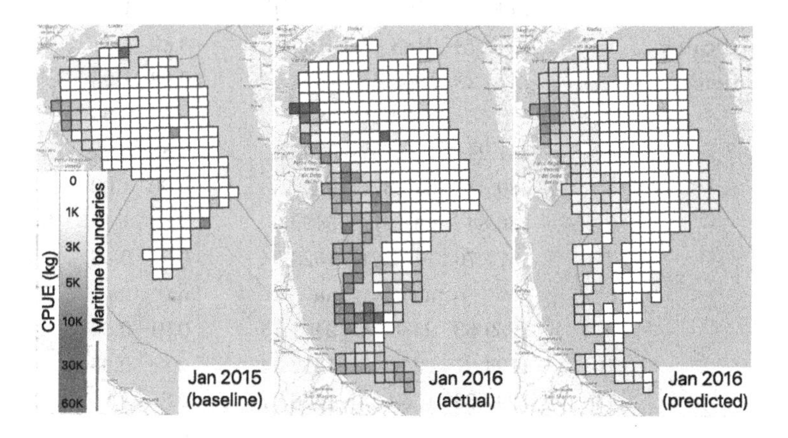

Fig. 3. CPUE over grid cells for January. Actual values for 2015 are the baseline for Jan. 2016 (left). Actual values for Jan. 2016 (middle) are used in the evaluation of values predicted by our model for Jan. 2016 (right).

Results and Discussion. The RAE values in Table 1 are less than 1 for every period, indicating that the RF model consistently performs better than the baseline. RAE on the last row for both models show a 13% improvement compared to baseline for the average results over all months. This naive model performance is largely due to the limited temporal extent of the data as described below, and also because the model is unaware of the spatial and temporal autocorrelation in the data. Incorporating the autocorrelations into the model is a subject for future work.

Table 1. Model evaluation metrics for monthly CPUE

	Month	MAE		CPUE mean	nMAE		RAE
	(2016)	RF	baseline	(actual)	RF	baseline	
Uniform catch distribution	1	2065.28	2473.68	2423.60	0.85	1.02	0.83
	2	1758.83	2473.26	1728.77	1.02	1.43	0.71
	3	812.98	881.96	983.87	0.83	0.90	0.92
	4	622.05	663.76	732.72	0.85	0.91	0.94
	5	862.11	948.41	936.90	0.92	1.01	0.91
	6	675.91	886.13	815.72	0.83	1.09	0.76
	7	2333.37	2377.42	2419.54	0.96	0.98	0.98
	8†	na	na	na	na	na	na
	9	3078.92	3168.13	3379.25	0.91	0.94	0.97
	10	1430.51	1705.33	1733.98	0.82	0.98	0.84
	11	2101.59	2295.61	2372.60	0.89	0.97	0.92
	12	1113.45	1213.80	1237.16	0.90	0.98	0.92
	All	1490.18	1707.45	1658.45	0.90	1.03	0.87
Weighted catch distribution	1	1947.78	2188.87	2193.30	0.89	1.00	0.89
	2	1474.28	1963.45	1593.04	0.93	1.23	0.75
	3	658.65	740.02	782.88	0.84	0.95	0.89
	4	441.24	486.75	529.31	0.83	0.92	0.91
	5	580.32	643.32	641.55	0.90	1.00	0.90
	6	436.84	637.21	542.09	0.81	1.18	0.69
	7	1474.76	1491.21	1523.94	0.97	0.98	0.99
	8†	na	na	na	na	na	na
	9	2239.63	2336.55	2461.78	0.91	0.95	0.96
	10	1124.27	1242.47	1312.44	0.86	0.95	0.90
	11	1297.63	1660.43	1526.47	0.85	1.09	0.78
	12	778.46	876.93	868.05	0.90	1.01	0.89
	All	1116.69	1290.72	1254.27	0.89	1.03	0.87

† No data available due to the fishing ban.

Figure 3 shows the baseline, actual, and predicted monthly CPUE for January. Presence of a grid cell on the map indicates that fishing activity occurred at least once during the whole month, and vice versa. Comparing the actual data for 2015 and 2016 in the figure, it is obvious that the fished area in January of 2016 is much larger than 2015. Since the model only uses the 2015 data for making predictions for 2016, it has no information about the areas that were not fished in that period of 2015. This presents a limitation which is imposed on the model due to the temporal restriction of the data. In other words, the model performance can be improved by having data for a longer period of time.

5 Conclusion and Future Work

In this paper, we explored a spatio-temporal dataset resulting from the fusion of the trajectories of the fishing vessels of the Northern Adriatic sea for 2015 and 2016, the landing report of the primary fish market of the area, and relevant environmental data. The landing reports represent quite a unique semantic feature to be associated with the fishing trajectory. Also, the utilization of environmental attributes influencing the species distribution (sea surface temperature, chlorophyll-a) and the fishermen behaviors (wave heights), represent an additional key element. Moreover, the utilization of the sea surface temperature can be helpful to evaluate the effect of climate changes on the fishing activities, a current and heavy issue needed to be addressed.

We applied to the dataset the Random Forest Machine Learning method, with the goal of predicting the CPUE indicator that could be helpful to improve the fisheries management plans for sustainable exploitation of the fishing resources. The forecast results—surpassing the baseline prediction by approximately 13%—indicate the value of the use of Machine Learning for this task, while clearly leaving a lot of room for improvement. Firstly, the task itself is difficult, and clearly a number of variables not currently captured—both environmental and latent (e.g., economic) factors, as well as fishermen behavior—influence the capabilities of the prediction model. On the one hand, this is due to the short temporal horizon of the landing and AIS data: two years, one for training and the other one for testing, are not sufficient. We expect that with access to 2017 data, the results will improve. On the other hand, other prediction techniques, such as the use of lag variables [21], or an alternative approach using modern time series prediction, should be exploited as we continue the project.

In data modeling, more sophisticated methods of catch distribution into grid cells, such as habitat selection [3], need to be looked at. A combination of both data modeling and machine learning extensions to this early research could also help to overcome the difficulties deriving from the short time period.

References

1. Armstrong, J.S., Collopy, F.: Error measures for generalizing about forecasting methods: empirical comparisons. Int. J. Forecast. **8**(1), 69–80 (1992)

2. Bogorny, V., Renso, C., de Aquino, A.R., de Lucca Siqueira, F., Alvares, L.O.: Constant–a conceptual data model for semantic trajectories of moving objects. Trans. GIS **18**(1), 66–88 (2014)
3. Bonanno, A., et al.: Habitat selection response of small pelagic fish in different environments. Two examples from the oligotrophic Mediterranean Sea. PLoS ONE **9**(7), e101498 (2014)
4. Breiman, L.: Bagging predictors. Mach. Learn. **24**(2), 123–140 (1996)
5. Breiman, L.: Random forests. Mach. Learn. **45**(1), 5–32 (2001)
6. Brüggemann, S., Bereta, K., Xiao, G., Koubarakis, M.: Ontology-based data access for maritime security. In: Sack, H., Blomqvist, E., d'Aquin, M., Ghidini, C., Ponzetto, S.P., Lange, C. (eds.) ESWC 2016. LNCS, vol. 9678, pp. 741–757. Springer, Cham (2016). https://doi.org/10.1007/978-3-319-34129-3_45
7. Copernicus: Europe's eyes on Earth. https://www.copernicus.eu/en
8. Dietterich, T.G.: Ensemble methods in machine learning. In: Kittler, J., Roli, F. (eds.) MCS 2000. LNCS, vol. 1857, pp. 1–15. Springer, Heidelberg (2000). https://doi.org/10.1007/3-540-45014-9_1
9. Estrada, J., Silva, C., Yáñez, E., Rodriguez, N., Pulido-Calvo, I.: Monthly catch forecasting of anchovy *Engraulis ringens* in the north area of Chile: non-linear univariate approach. Fish. Res. **86**(2), 188–200 (2007)
10. Ferrà, C., et al.: Mapping change in bottom trawling activity in the Mediterranean Sea through AIS data. Mar. Policy **94**, 275–281 (2018)
11. Hyndman, R.J., Athanasopoulos, G.: Forecasting: Principles and Practice, 2nd edn. OTexts, Melbourne (2018)
12. Maina, I., Kavadas, S., Somarakis, S., Tserpes, G., Stratis, G.: A methodological approach to identify fishing grounds: a case study on Greek trawlers. Fish. Res. **183**, 326–339 (2016)
13. Mello, R.d.S., et al.: MASTER: a multiple aspect view on trajectories. Trans. GIS (2019, to appear). https://doi.org/10.1111/tgis.12526
14. Mills, C.M., Townsend, S.E., Jennings, S., Eastwood, P.D., Houghton, C.A.: Estimating high resolution trawl fishing effort from satellite-based vessel monitoring system data. ICES J. Mar. Sci. **64**(2), 248–255 (2006)
15. Nurdin, S., Ahmad Mustapha, M., Lihan, T., Ghaffar, M.A.: Determination of potential fishing grounds of *Rastrelliger kanagurta* using satellite remote sensing and GIS technique. Sains Malays. **44**(2), 225–232 (2015)
16. Paige Eveson, J., Hobday, A., Hartog, J., Spillman, C., Rough, K.: Seasonal forecasting of tuna habitat in the Great Australian Bight. Fish. Res. **170**, 39–49 (2015)
17. Parent, C., et al.: Semantic trajectories modeling and analysis. ACM Comput. Surv. (CSUR) **45**(4), 42 (2013)
18. Soares, A., et al.: CRISIS: integrating AIS and ocean data streams using semantic web standards for event detection. In: International Conference on Military Communications and Information Systems (2019)
19. Soares Júnior, A., Moreno, B.N., Times, V.C., Matwin, S., Cabral, L.d.A.F.: GRASP-UTS: an algorithm for unsupervised trajectory segmentation. Int. J. Geogr. Inf. Sci. **29**(1), 46–68 (2015)
20. de Souza, E.N., Boerder, K., Matwin, S., Worm, B.: Improving fishing pattern detection from satellite AIS using data mining and machine learning. PLoS ONE **11**(7), e0158248 (2016)
21. Tyralis, H., Papacharalampous, G.: Variable selection in time series forecasting using random forests. Algorithms **10**, 114 (2017)
22. Vespe, M., Gibin, M., Alessandrini, A., Natale, F., Mazzarella, F., Osio, G.C.: Mapping EU fishing activities using ship tracking data. J. Maps **12**, 520–525 (2016)

BY

A Neighborhood-Augmented LSTM Model for Taxi-Passenger Demand Prediction

Tai Le Quy[1], Wolfgang Nejdl[1](✉), Myra Spiliopoulou[2], and Eirini Ntoutsi[1]

[1] L3S Research Center, LUH, Hannover, Germany
{tai,nejdl,ntoutsi}@l3s.de
[2] Otto-von-Guericke-University, Magdeburg, Germany
myra@ovgu.de

Abstract. Taxi is a convenient means of transportation worldwide. Accurately predicting the taxi-demand is crucial for taxi-companies to effectively allocate their fleet to taxi-stands and reduce the waiting time for passengers thus increasing their overall satisfaction and customer retention. Nowadays precise information about taxi-rides is available and can be used to infer the taxi-passenger demand across different locations and time-points. In this paper, we propose an approach for predicting the pick-demand of a given taxi-stand, that takes into account not only the demand-history of the particular stand but it also considers information from neighboring stands. Our model is an LSTM neural network augmented with information from the spatial neighborhood of the stands. Experiments with two versions of the taxi demand dataset from the city of Porto, Portugal show that our approach can provide better predictions comparing to approaches that do not exploit the neighborhood.

Keywords: Taxi-passenger demand · Time series prediction · LSTM · k-nearest neighbors · Deep learning · Neural networks

1 Introduction

Advances in sensor and wireless communication contribute to the development of intelligent transportation systems, which lead to the transformation of transportation domain. Taxi networks are the important means of transportation providing the convenient and direct services for passengers. Currently, many taxi vehicles are equipped with Global Positioning System (GPS) and wireless communication features that can generate a new source of rich spatial temporal information.

Intelligent online systems that plays a crucial role for real time taxi services scheduling, taxi sharing, fuel-saving routing, time-saving route finding are already developed to improve taxi service reliability [19]. Improving levels of passenger satisfaction and maximal profit for taxi providers are the main targets of taxi companies. Balancing the relationship between the passenger demand and the number of running taxi vehicles is the most efficient way to maximize the profit for taxi providers [19]. Knowledge on time and places that is emerged

K. Tserpes et al. (Eds.): MASTER 2019, LNAI 11889, pp. 100–116, 2020.
https://doi.org/10.1007/978-3-030-38081-6_8

the passenger demand can be an advantage for drivers even when there is no economic availability. The information regarding passenger demand is very useful for drivers in making decision moving to pick up passengers in a particular region in the city. GPS historical data are the main variable used in prediction models because it can reveal hidden mobility patterns.

Many researchers were attracted by the mobility data and proposed different approaches for taxi-passenger demand prediction. Among the investigated models are linear regression, ARIMA, feed-forward neural networks and more recently, deep neural networks. Most of these approaches focus exclusively on the information of the stand to predict passenger demand in the future. In this paper, we also exploit information from neighboring stands in an attempt to enrich the information provided to the model, in our case an LSTM neural network. Data augmentation [26] is a popular technique especially for data-insensive models like Deep Neural Networks (DNNs); for example, the improved performance on ImageNet [3] was also attributed to image augmentation using different domain-specific augmentation techniques like image reflection, translation, cropping and changing the color palette. Unnikrishnan et al. [25] introduce an entity-centric stream classification approach that exploits the observation history of the particular entity and of entities similar to it. Similar entities are defined on the basis of static entity characteristics like gender and birthdate in case of patient data or product properties in case of product reviews. Our neighborhood-selection idea is similar as we also rely on the spatial neighborhood of the taxi stands rather than their demand histories. As our experiments with data from the taxi network in the city of Porto, Portugal spanning a period of one year show, such an augmentation is beneficial for the predictive performance of the model.

The rest of the paper is organized as follows: Sect. 2 overviews the related work. Our neighborhood-augmented LSTM approach is presented in Sect. 3. A detailed experimental evaluation is provided in Sect. 4. Finally, conclusions and outlook are summarized in Sect. 5.

2 Related Work

There is a large body of work on traffic-related data, from trajectory querying, to hotspot detection, clustering, trajectory prediction [1] etc. Hereafter, we focus mainly on existing works using taxi-data and related mainly to our demand prediction problem.

A taxi-sharing framework is proposed in [7] that returns the top-k taxi recommendations for a passenger request. They select the *top–k* candidate taxis for a specific location by considering its neighbors on the traffic network. For their experiments, they have used the New York city taxi dataset. Luca *et al.* [6] proposes a method to find the Nash equilibrium in a taxi sharing fare in case there are many passengers sharing one taxi in order to save money. For their experiments, they also use the New York city dataset.

The problem of taxi-passenger demand prediction has attracted the attention of many researchers recently and as result, several approaches have been proposed. Most of these approaches rely on well-known prediction models from the time-series forecasting domain [15]. Kaltenbrunner et al. [11] introduced an auto-regressive moving average (ARMA) model approach to forecast the number of bicycles at a station from Barcelona's bicycle network in order to increase the stations spatial deployment. Min and Wynter [17] applied another popular time-series prediction model, ARIMA (Auto-Regressive Integrated Moving Average) to predict the speed and volume of traffic in a road network. Luis *et al.* [18,19] introduce an ensemble of experts to predict taxi demand, where each expert is specialized to a particular trend. In particular, their ensemble consists of a Time-Varying Poisson model, a Weighted Time-Varying Poisson model and a ARIMA model. The experiments were conducted on the Porto taxi dataset. Su *et al.* [27] predicts taxi-passenger demand in urban areas in Hong Kong using multiple features such as the number of vacant taxi on the roads, the waiting time, passenger demand, taxi fare as the input for a feed-forward neural network. Recently, TONG, Yongxin *et al.* [24] presented a multi-dimensional linear regression model to predict the taxi demand in Beijing and Hangzhou, China. Their multi-dimensional representation consists of temporal features, spatial features, meteorological features, and the combination of these features. Yao et al. [28] proposed a deep learning framework to model both spatial and temporal relations by using two neural network model CNN and LSTM to predict taxi demand in Guangzhou, China.

Contrary to most of the existing works that rely exclusively on taxi-stand's own demand history we enrich the data representation of each stand using information from neighboring stands. Our intuition is that the demand of a taxi-stand might be indicative of the demand of some nearby stand as well. Such an augmentation is especially beneficial for data intensive models, our base model is an LSTM deep neural network model, in order to reduce over-fitting and eventually, generalization performance.

3 Neighborhood-Augmented Taxi Demand Prediction

3.1 Problem Definition

Let $S = \{s_1, s_2, .., s_N\}$ be the set of predefined N taxi-stands in a city. Consider $X_s = \{X_{s,0}, X_{s,1}, .., X_{s,t}\}$ to be a discrete time series (based on an aggregation period of P-minutes) that models the taxi-demand for stand s, that is, the number of pick-ups for each aggregation period P at s. We refer to this time series as the *demand history of stand s*. Our goal is to build a model which predicts the demand $X_{s,t+1}$ for the next time point $t+1$ at taxi-stand s.

Traditional approaches rely solely on the demand history of the stand X_s for the prediction (we use such methods as baselines for our comparison, c.f., Sect. 4.3). In this work we propose to augment the stand's demand history X_s with information from its neighborhood. The intuition behind this augmentation process is that nearby taxi-stands might display similar demands. Our dataset seems to justify our intuition: In Fig. 1 we show the spatial proximity of the different taxi-stands (left) vs their demand proximity (right). The demand proximity is evaluated using Pearson correlation and for efficiency reasons, only part of the history demand. Due to space, we show here only the information for the first 20 taxi-stands (IDs 1–20),

As we can see, when the pairwise spatial distances are high, an opposite trend is observed in the demand history correlation values. This can be observed for a variety of taxi-stands, for example, 4, 5 and 8.

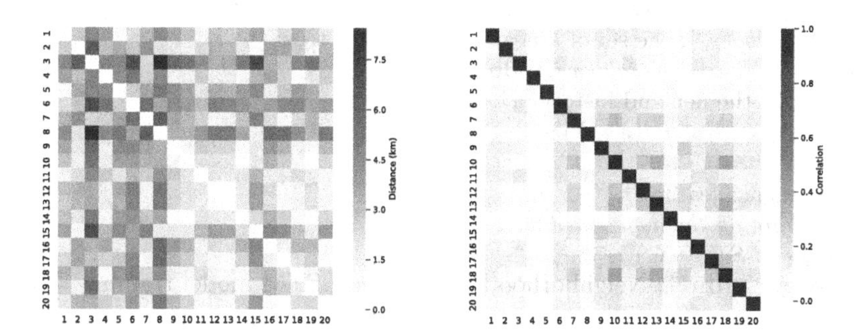

Fig. 1. Spatial proximity (left) vs pickup demand correlation (right) between taxi-stands (based on dataset D1).

Based on this motivation, we present hereafter the neighborhood-augmented LSTM model for predicting taxi-passenger demand of a taxi-stand.

3.2 Neighborhood-Augmented LSTM Model

Our model is an extension of the well known Long Short-Term Memory (LSTM) networks [8], a special kind of a recurrent neural network (RNN). A common LSTM unit is composed of a cell, an input gate, an output gate and a forget gate. The cell remembers values over arbitrary time intervals and the three gates regulate the flow of information into and out of the cell. There are several architectures of LSTM units. An LSTM cell takes an input and stores it for some period of time. Because the derivative of the identity function is constant, when an LSTM network is trained with back propagation through time, the gradient does not vanish. The activation function of the LSTM gates is often the logistic function. Intuitively, the input gate controls the extent to which a new value flows into the cell, the forget gate controls the extent to which a value remains in the cell and the output gate controls the extent to which the value in the cell is used to compute the output activation of the LSTM unit. There are connections into and out of the LSTM gates, a few of which are recurrent. The weights of these connections, which need to be learned during training, determine how the gates operate.

In our approach, we train an LSTM model for each taxi-stand s using not only its primary demand history X_s but also demand history information from its k-nearest neighbors. That is, the input to the LSTM is a $(k+1)$ dimensional vector, X'_s. The actual demand values (ground truth) comes from taxi-stand s and therefore the goal is to fit the neighborhood-augmented LSTM model for predicting the demand values of taxi-stand s.

input : Taxi demand dataset; k-number of neighbors
output: Prediction model M_s for taxi-stand s

1 *//Data augmentation*
2 X_s: the demand history of taxi-stand s up to time t
3 $X'_s \leftarrow X_s$ *//extended representation*
4 $\{Neighbors_s\}$: the set of k nearest taxi-stands to s
5 **for** $i \leftarrow 1$ **to** $|\{Neighbors_s\}|$ **do**
6 | X_i: the demand history of taxi-stand i
7 | $X'_s \leftarrow Extend(X'_s, X_i)$
8 **end**
9 Normalize features
10 *//Train on the augmented data*
11 $M_s \leftarrow LSTM(X'_s)$

Algorithm 1: Neighborhood-augmented LSTM model training

The pseudo code of the algorithm is shown in Algorithm 1. Each taxi -stand has it own LSTM model for training.

In the above algorithm, the normalization step aims to normalize all features in the [0–1] range. This is an important step for LSTM convergence [13]. In particular, we use min-max normalization.

The structure of our LSTM network is shown in Fig. 2 and explained hereafter. In this model, time series of stand X with its $k - neighbors$ are used as the input of the first LSTM layer, followed by a hidden layer before a dropout unit. Predicted time series Y is the result of our model. The tuning of the hyper-parameters is discussed in detail in Sect. 4.4 but the selected values are mentioned here as well:

1. Input (X'_s, the extended description of stand s; look back value = 5 (see Sect. 4.4.))
2. LSTM (N=200, optimizer = 'Adamax', Activation function = 'tanh', loss= 'mean squared error', batch size = 100 (see Sect. 4.4.))
3. Full connected LSTM (N=200, Activation function ='tanh')
4. Dropout = 0.7 (see Sect. 4.4.)
5. Dense (Activation function = 'tanh')

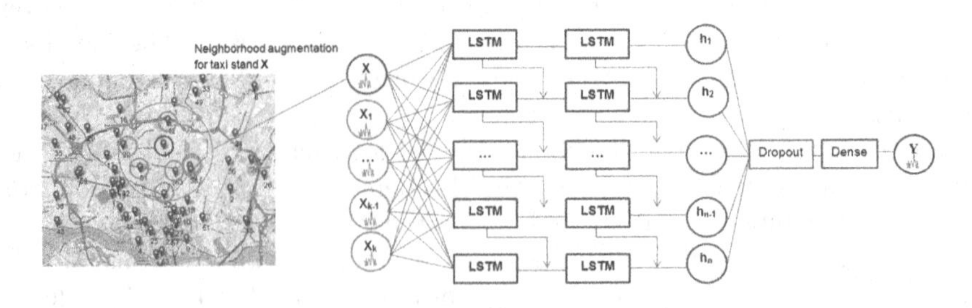

Fig. 2. The architecture of the neighborhood-augmented LSTM.

4 Experimental Evaluation

We evaluate our approach on the publicly available dataset on taxi-demand from Porto, Portugal (Sect. 4.1). The experimental setup and evaluation criteria are discussed in Sect. 4.2. The goal of our experiments is to evaluate the impact of the neighborhood-based augmentation on prediction quality (Sect. 4.5) as well as to study how the "quality" of the neighborhood as evaluated by the average distance of neighboring stands from the reference stand affects the predictions (Sect. 4.6).

Our LSTM approach was implemented using Keras and Tensorflow, whereas for the other approaches we use the available implementations in Python.[1].

4.1 Dataset

We use the dataset from [19] that contains information on taxi trips organized by a taxi company in the city of Porto in Portugal and was part of the ECML 2015 challenge[2]. The dataset spans over a period of one year (from July 2013 to June 2014) and contains 1.710.670 records. Each record corresponds to a completed taxi trip, described in terms of 9 features: (1) TRIP_ID: a unique identifier for each trip; (2) CALL_TYPE: the way to use the taxi service and contain one of three possible values: A (the trip is assigned from the call central), B (the trip is departed from a specific stand) or C (passengers are pickup from a random street); (3) ORIGIN_CALL; (4) ORIGIN_STAND: a unique identifier for the taxi-stand; (5) TAXI_ID; (6) TIMESTAMP: Unix Timestamp (in seconds); (7) DAYTYPE: the daytype of the trip's start (holiday or any other special day); (8) MISSING_DATA; (9) POLYLINE: the trajectory of trip. In addition, the dataset also provides the information of all 63 taxi-stands with their name and GPS coordinates.

Figure 3 depicts the spatial distribution of the taxi-stands in Porto, each stand is assigned a unique ID from 1 to 63. As one can see, the stands are not randomly distributed rather their spatial density reflects the demand with most stands located close to the city center. Moreover, we can see that despite the aforementioned mandatory regulation there are trips that do not start at the location of the taxi-stand. The intensity of the color in Fig. 3 shows the density of the starting points; in many cases taxi-stands have the highest local density but not all trips start at some taxi-stand.

We preprocess the data as follows: Firstly, we sort all records by timestamp in ascending order. We remove features MISSING_DATA and POLYLINE and we add two new features: LATITUDE and LONGITUDE extracted from the POLYLINE attribute and describing the coordinates of the starting trip location. After, we remove instances that have no both taxi-stand ID starting location. This results in a clean dataset of 1.706.572 completed taxi trips. Contrary to previous work [18,19], we create two versions of the dataset for the experiments.

[1] The code is available at: https://github.com/quytai3985/PortoTaxiPrediction.
[2] https://www.kaggle.com/c/pkdd-15-predict-taxi-service-trajectory-i/data.

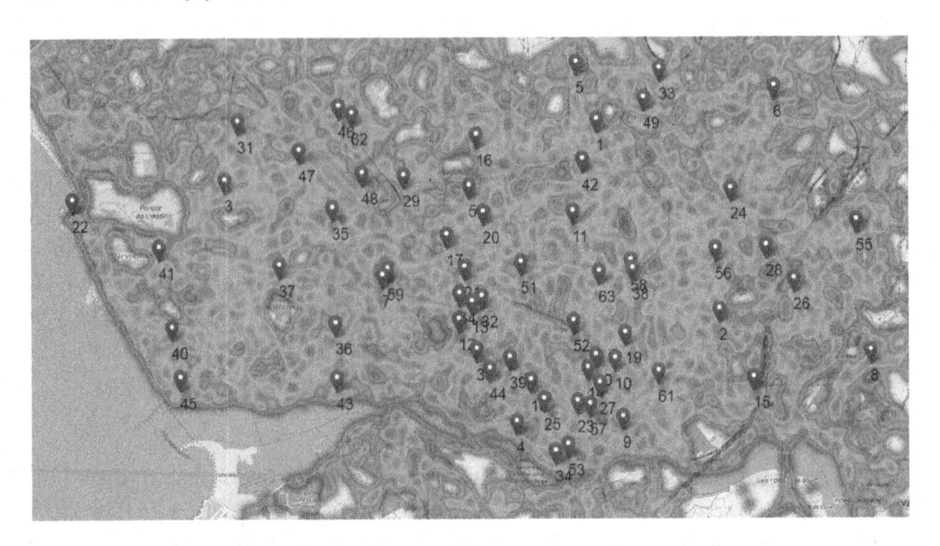

Fig. 3. Spatial distribution of the taxi-stands. Numbers 1–63 indicate the IDs of the stands.

The first dataset (D1) has only the taxi trips with CALL_TYPE equal to 'B', i.e., all trips that are departing from some taxi-stand. This dataset contains 817.861 instances and can be used for building a prediction model that can forecast the short term demand for specific taxi-stands.

However, as already mentioned not all trips start from a taxi-stand (i.e., the initial location does not match the location of some taxi-stand). Due to the amount of these trips (888.711 trips, 52.07% of the overall dataset), this information cannot be easily omitted, rather these trips might play an important role for the predictions and one needs to consider them for the forecasting. Therefore, in the second version (D2), we use all records of the clean dataset. For the trips that do not start from a taxi-stand (i.e., those with a CALL_TYPE equal to 'A' or 'C'), we assign them to their closest taxi-stand based on distance between the starting location of the trip and the location of the taxi-stand. Intuitively, we consider the taxi-stand as covering some region in the city with the its coordination being the center of this region.

The distribution of the trips on the different taxi-stands for the (D1), (D2) datasets is shown in Figs. 4 and 5, respectively. For each dataset, we also display the mean and median demand values. It is easy to observe that there are a large number of taxi demand in several stands. For example, the most popular stand is stand 15, which corresponds to the main train station. The top 10 most crowded stands in **D1** account for approximately 46.5% of the total 817.861 passenger demand. In dataset **D2**, that proportion is around 36.3% of 1.7M pickups. A closer look at the top 10 stands via Google Maps reveals that they are all close to the main train station and the city center with many historical sites, shops and hotels.

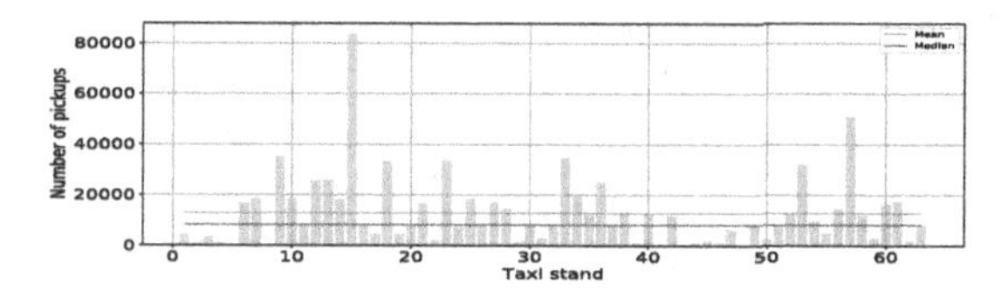

Fig. 4. Pickup distribution per taxi-stand on D1.

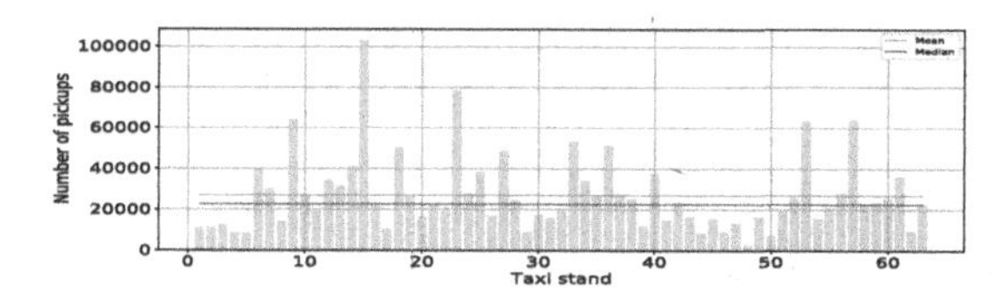

Fig. 5. Pickup distribution per taxi-stand on D2.

4.2 Experimental Setup and Evaluation Measures

We set the aggregation period at 30 min based on the average waiting time at a taxi-stand as in [19]. We generate the demand history at each taxi-stand by aggregating the number of pick ups every 30 min.

Demand history examples are presented in Fig. 6 for taxi-stands 1 and its spatial neighbor - taxi-stand 49 during one week (from 01 Jul 2013 to 07 Jun 2013). Similarly, Fig. 7 describes the demand history of taxi-stand 15 and its spatial neighbor, taxi-stand 61. In both figures, we use D1 as the data source. As we can see, the pickup demand series are different and depended on the location of the taxi-stand. For example, taxi-stand 1 is located far from the city center (around 5 km) whereas taxi-stand 15 is close to the main train station. As a result, the demand on stand 15 is much higher with over 83.000 yearly pickups whereas the demand for taxi-stand 1 is only 4.500. Similarly, the taxi demand on stand 61 (close to stand 15) is around 17.000 pickups whereas the demand for taxi-stand 49 (close to stand 1) is only 8.000. Except for the differences in the amplitude of the demand, we can also see differences w.r.t. the temporality of the demand. For example, both stands 15 and 61 have around-the-clock demand, but this is not the case for stands 1 and 49. This behavior comprises our motivation behind the proposed neighborhood-augmented demand prediction model.

In time series prediction, the measurement symmetric Mean Absolute Percentage Error (sMAPE) [16] is more meaningful than other measurement, such as MSE, RMSE. One reason is the proportion values are more comprehensive than squared errors [21]. As a consequence, in our experiment, we evaluate the prediction quality of the models for each taxi-stand by comparing the forecast values with the original ones using sMAPE. However, we still report our results on MSE measurement as a reference one. In particular, let the true demand for a taxi-stand s be: $X_s = \{X_{s,0}, X_{s,1}, .., X_{s,t}\}$ and the predicted demand:

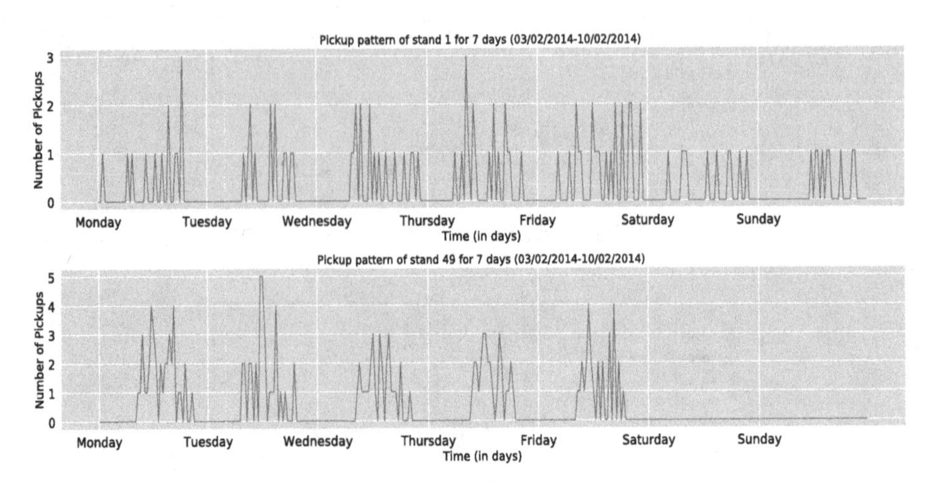

Fig. 6. Pickup demand history for nearby taxi-stands 1 and 49.

$\hat{X}_s = \{\hat{X}_{s,0}, \hat{X}_{s,1}, .., \hat{X}_{s,t}\}$. Then $sMAPE_s$ is given by:

$$sMAPE_s = \frac{100\%}{t} \sum_{i=1}^{t} \frac{\mid X_{s,i} - \hat{X}_{s,i} \mid}{(X_{s,i} + \hat{X}_{s,i})/2} \tag{1}$$

In Eq. 1, sMAPE values fluctuate between −200% and 200% [16]. Flores [5] claims that a percentage error between 0% and 100% is much easier to interpret, therefore we omit factor 2 in the denominator. Furthermore, due to possible prediction of negative demand values, we use absolute values in the denominator of Eq. 1. Additionally, Eq. 1 can result in a high error if the real demand is 0 and the predicted one is non-zero; in such a case, the error would be 100%. To deal with this issue, we use *Laplace correction* [10] by adding a constant c to the denominator. Finally, the modified $sMAPE_s$ that is used for our evaluation is given by:

$$sMAPE_s = \frac{100\%}{t} \sum_{i=1}^{t} \frac{\mid X_{s,i} - \hat{X}_{s,i} \mid}{\mid X_{s,i} \mid + \mid \hat{X}_{s,i} \mid + c} \tag{2}$$

The constant c is user-defined. In our experiments, we use the corrected sMAPE version (Eq. 2) with $c = 1$. The aforementioned formulas refer to the error at each stand, we aggregate the error over all taxi-stands as follows:

$$sMAPE = \frac{\sum_{i=1}^{N} sMAPE_i}{N} \tag{3}$$

where N is the number of taxi-stands.

4.3 Baselines and Method Parameter Settings

We compare our approach against well-known prediction methods, described hereafter together with their parameter tuning.

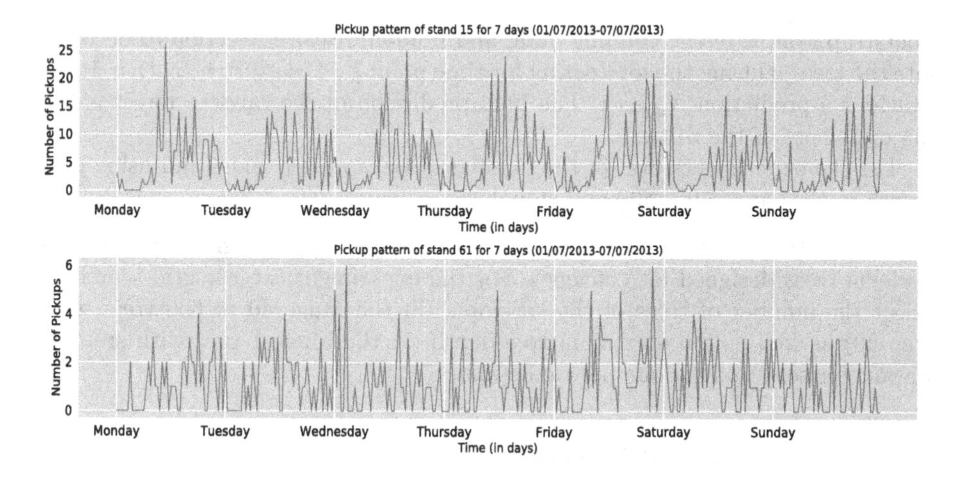

Fig. 7. Pickup demand history for nearby taxi-stands 15 and 61.

Simple Moving Average: A simple moving average (SMA) [9] is an arithmetic moving average calculated by averaging the observed values of a time series in the calculation period. Given a calculation period of q timepoints, the prediction $X_{s,t+1}$ for the next time point $t+1$ is given by:

$$X_{s,t+1} = \frac{1}{q+1} \sum_{j=0}^{q} X_{s,t-j} \tag{4}$$

The number of periods q should be set; when $q = 0$ this is simply the value of the last observation. For our experiments, we choose $q = 20$ using grid search. We set the range of q from 2 to 24 (equals to 1–12 h) with step 1. The selection of q was based on taxi-stand 1 and dataset `D1`. Taxi-stand 1 is chosen as the representative stand for tuning as its location is far from the places that concentrate a huge amount of vehicles, such as the main station or city center. Moreover, in our experiments the performance of the different models on this taxi-stand was close to the average values.

Linear Regression: In a linear regression model [22] the future value of a variable is assumed to be a linear function of its past q values, where q defines the amount of past values contributing to the computation.

$$X_{s,t+1} = \beta_0 + \beta_1 X_{s,t-1} + \beta_2 X_{s,t-2} + .. + \beta_q X_{s,t-q} \tag{5}$$

For our experiments, we choose $q = 15$ using grid search, similarly to parameter selection for SMA. We apply $q = 15$ for all 63 different models/taxi-stands. However, the parameter β_0 is adapted to each taxi-stand using grid search with β_0 in the range 10^{-16} to 10^6 and step 100.

Random Forest Regression: Random forest [14] is an ensemble technique averaging the forecasting of a large number of decorrelated decision trees. Random forests are built on two main ideas - bagging to build each tree on a different bootstrap sample of the training data, and random feature selection to decorrelate the trees. During the forecasting for time point $t+1$, each tree B_j $(j = 1..m)$ provides a prediction $X_{s,t+1,j}$. The final prediction of the random forest is the majority vote of the m trees.

For the experiments, we set the number of trees m per taxi-stand using grid search in the range $10 - 800$ and step 40.

XGBoost Regression: XGBoost [2] is an implementation of gradient boosting decision trees designed for efficiency. For the experiment, we use grid search to select the number of trees of the ensemble (in the range 40 to 600 trees with step 40) as well as the maximum tree depth (in the range 1 to 4 with step 1). Parameter selection is done per taxi-stand.

4.4 LSTM Parameter Settings

The number of neighbors k is selected by grid search based on the representative taxi-stand 1; the result is a value of $k = 15$. A similar process is followed for the rest of the parameters, i.e., they were set using grid search over the data from the representative stand 1. In particular, the *look back value* parameter is select from a range of 2 to 24 (corresponding to 1 to 12 h in the history) with step 1. The best *look back value = 5* is chosen as it raises best value of $sMAPE$. *AdamMax* and *tanh* are selected for the gradient descent optimization algorithm and activation function, respectively as they cause the best $sMAPE$ values compared to other functions. Additionally, a list of possible candidates (10, 15,20,25,50,100,200,500 1000) is investigated to find the optimal *epoch* and *batch size* number. The best results were obtained with *epoch=25* and *batch size = 100.* Furthermore, the range from 10 to 300 with step 10 and the range from 1 to 4 with step 1 were explored to find the best number neurons per layer and the number of hidden layers, respectively. According to the results, we construct our model with 1 hidden layer and $N = 200$ neurons. Besides, to prevent our LSTM model from *overfitting* we use the *dropout* technique that randomly drop units (along with their connections) from the neural network during training in order to avoid co-adapting too much [23]. The dropout rate was set to 0.7, base on our experiments with a range of dropout values fro 0.1 to 0.9 with step 0.1.

4.5 Taxi-Demand Prediction Quality Results

Table 1 summarizes the prediction quality of the different models for dataset `D1`, containing trips starting from an actual taxi-stand. In this table, neighborhood-augmented LSTM is experimented with $k = 15$. Table 2 summarizes the results for dataset `D2`., containing all trips from the cleaned dataset and after mapping the trips that do not start from a stand to their closest stand. $k = 25$ is the number of neighbors used in Neighborhood-augmented LSTM architecture.

As we can see, our approach, Neighborhood-augmented LSTM, results in the smallest sMAPE errors, followed by vanilla LSTM. Moreover, the LSTM models outperform traditional prediction models with linear regression models

Table 1. Prediction quality of the different models on D1.

Model	Training		Testing	
	sMAPE (%)	MSE	sMAPE (%)	MSE
Simple moving average			23.34	1.721
Linear regression	24.37	1.61	24.52	1.765
Random forest regression	**16.83**	**0.383**	24.25	1.660
XGBoost regression	23.90	1.391	23.91	**1.585**
LSTM	18.37	1.659	18.54	1.839
Neighborhood-augmented LSTM	17.32	1.465	**17.63**	1.682

Table 2. Prediction quality of the different models on D2

Model	Training		Testing	
	sMAPE (%)	MSE	sMAPE (%)	MSE
Simple Moving Average			30.33	**5.369**
Linear Regression	30.78	4.206	31.23	5.988
Random Forest Regression	**18.49**	**0.715**	31.03	5.503
XGBoost Regression	30.466	3.605	30.51	5.449
LSTM	27.03	4.16	27.22	6.660
Neighborhood-augmented LSTM	25.88	3.84	**26.07**	6.444

performing worse in both datasets. The improvement rates are higher for dataset D1 comparing to dataset D2. A possible reason is the assignment of the trips to their closest taxi-stands, a process that might have introduced errors. We plan to investigate alternative assignments in our future work, for example some weight decay approach based on the distance of the pick-up from its closest taxi-stand or soft assignments to multiply nearby taxi-stands.

A closer look at the performance of our approach vs the original demand for the different taxi-stands is presented in Figs. 8 and 9 for datasets D1, D2, respectively. As we can see, two different patterns of performance are shown. In dataset D1, the performance of the model has large variation, probably due to the large deviation of pickups among taxi-stand. The picture is different in dataset D2, where the actual number of pickups appears more balanced across the taxi-stands.

The variation in the performance of the different prediction models over the different taxi-stands is demonstrated more clearly in Fig. 10, where each boxplot corresponds to one prediction method and summarizes the sMAPE error over all stands. As we can see, there is large variation in D1 for all methods. Moreover, traditional approaches like MSA, LR, RF and XGBoost display skewed performance whereas the LSTM approaches are symmetric so the error over the different stands follows a normal distribution. Interestingly, and despite the lower performance of the methods on dataset D2 comparing to D1, the spread of the error across the taxi-stands is very small for all methods, although there exist outliers. In case of LSTM-based models, most of the outliers correpond to stands with better predictions (lower sMAPE). Another interesting observation is that the model performs best when the number of pickups is close to average demand. As an extreme case, the most popular stand, stand 15 corresponding to the main train station, has the highest error on both datasets D1 and D2. A possible explanation is that such a stand is very difficult to model with a single model and one might need to consider different models for different contexts (e.g. season based, weekdays vs weekends etc). We leave this as our future work.

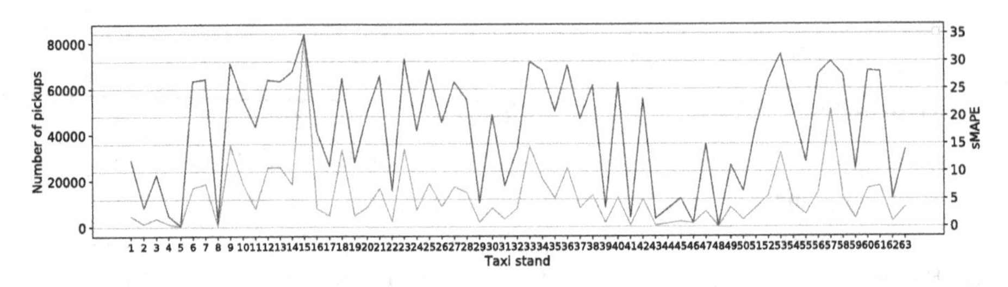

Fig. 8. Real demand vs neighborhood-augmented LSTM error across different taxi-stands for dataset D1.

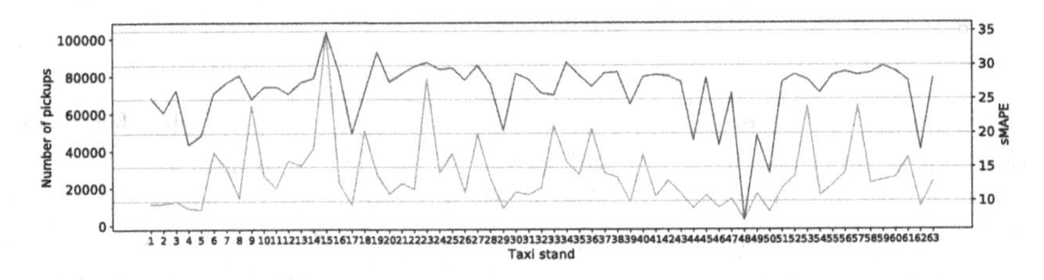

Fig. 9. Real demand vs neighborhood-augmented LSTM error across different taxi-stands for dataset D2.

4.6 Impact of Neighborhood

Our augmentation approach is based on the number of neighbors k parameter. We evaluate the impact of k on the predictive performance within a range of k from 1 to 61 and step 4. The results for both datasets D1 and D2 are shown in Figs. 11 and 12, respectively. The effect of k is more pronounced when testing with dataset D1. On dataset D1, when k is greater than 15 or the average distance from a specific taxi-stand to its neighbors is farther than 1 km, the performance of LSTM has a light fluctuation. While these values in dataset D2 are 25 and approximately 1.7 km, respectively. This shows that the proximity taxi-stands have a great influence on the prediction ability of the model. This is understandable because passengers in remote locations will be difficult to access the current pick-up stand for a short time.

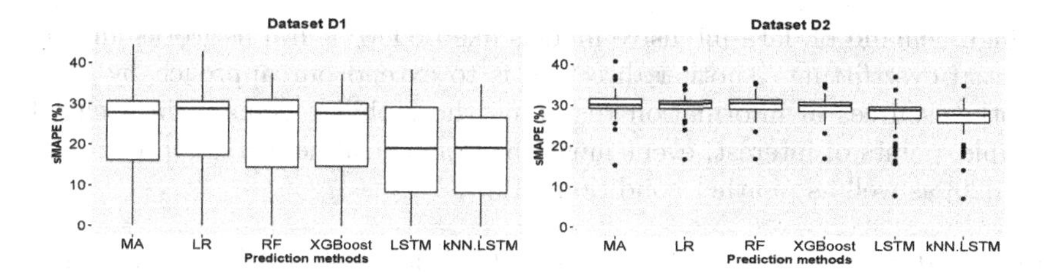

Fig. 10. Comparing error distributions for different prediction methods for dataset D1 (left) and D2 (right).

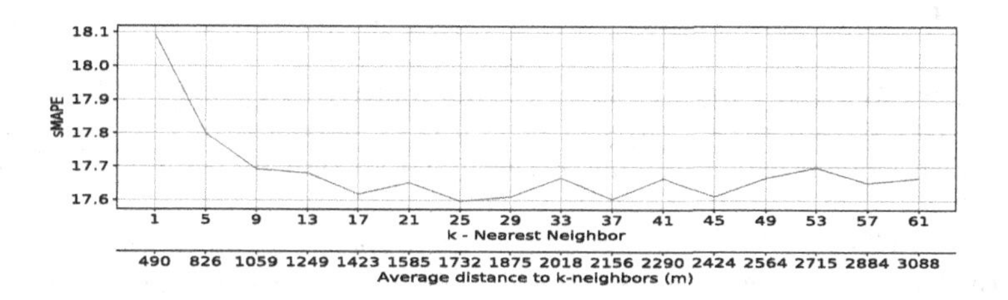

Fig. 11. Evaluating the impact of neighborhood on the predictive performance of neighborhood-augmented LSTM model on: D1

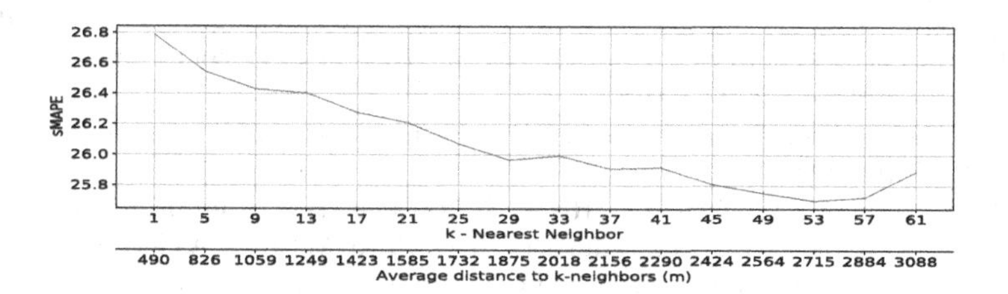

Fig. 12. Evaluating the impact of neighborhood on the predictive performance of neighborhood-augmented LSTM model on: D2

5 Conclusions and Outlook

In this paper we propose a neighborhood-augmented LSTM model for predicting the pick-demand of a given taxi-stand. Our experiments show that such an augmentation benefits the predictive performance of the model comparing to an LSTM approach that exploits strictly the demand history of the taxi-stand as well as to traditional prediction methods like SMA and regression.

There are several extension possibilities. In this work, we have considered a global neighborhood threshold k for all taxi-stands. However a more careful selection of the neighborhood and eventually a stand-tuned k would be more appropriate in order to account for different demand densities and taxi-stand densities in the city. Such a tuning could also take into account the data sparsity in the taxi-stand and grow the neighborhood progressively in order to cope with the high demand of data-intensive models like LSTM neural networks and their potential overtfitting. Another direction is to extend our approach by including other sources of information regarding the mobility demand in a city, for example, points of interest, event mentions from social networks [4], traffic patterns [20] as well as weather conditions [12].

Acknowledgement. The work was inspired by the German Research Foundation (DFG) project OS-CAR (Opinion Stream Classification with Ensembles and Active leaRners) for which the last two authors are Principal Investigators.

References

1. Alahi, A., Goel, K., Ramanathan, V., Robicquet, A., Fei-Fei, L., Savarese, S.: Social LSTM: human trajectory prediction in crowded spaces. In: Proceedings of the IEEE Conference on Computer Vision and Pattern Recognition, pp. 961–971 (2016)
2. Chen, T., Guestrin, C.: Xgboost: a scalable tree boosting system. In: Proceedings of the 22nd ACM Sigkdd International Conference on Knowledge Discovery and Data Mining, pp. 785–794. ACM (2016)
3. Deng, J., Dong, W., Socher, R., Li, L.J., Li, K., Fei-Fei, L.: Imagenet: a large-scale hierarchical image database. In: 2009 IEEE Conference on Computer Vision and Pattern Recognition, pp. 248–255. IEEE (2009)
4. Fafalios, P., Iosifidis, V., Stefanidis, K., Ntoutsi, E.: Tracking the history and evolution of entities: entity-centric temporal analysis of large social media archives. Int. J. Digital Libr. 1–13 (2018)
5. Flores, B.E.: A pragmatic view of accuracy measurement in forecasting. Omega **14**(2), 93–98 (1986)
6. Foti, L., Lin, J., Wolfson, O., Rishe, N.D.: The nash equilibrium among taxi ridesharing partners. In: Proceedings of the 25th ACM SIGSPATIAL International Conference on Advances in Geographic Information Systems, p. 72. ACM (2017)
7. Fu, X., Huang, J., Lu, H., Xu, J., Li, Y.: Top-k taxi recommendation in realtime social-aware ridesharing services. In: Gertz, M., et al. (eds.) SSTD 2017. LNCS, vol. 10411, pp. 221–241. Springer, Cham (2017). https://doi.org/10.1007/978-3-319-64367-0_12
8. Hochreiter, S., Schmidhuber, J.: Long short-term memory. Neural Comput. **9**(8), 1735–1780 (1997)
9. Hyndman, R.J.: Moving averages. International encyclopedia of statistical science pp. 866–869 (2011)
10. Jaynes, E.T.: Probability Theory: The Logic of Science. Cambridge University Press, Cambridge (2003)
11. Kaltenbrunner, A., Meza, R., Grivolla, J., Codina, J., Banchs, R.: Urban cycles and mobility patterns: exploring and predicting trends in a bicycle-based public transport system. Pervasive Mob. Comput. **6**(4), 455–466 (2010)
12. Koutroumanis, N., Santipantakis, G.M., Glenis, A., Doulkeridis, C., Vouros, G.A.: Integration of mobility data with weather information (2019)
13. Laurent, C., Pereyra, G., Brakel, P., Zhang, Y., Bengio, Y.: Batch normalized recurrent neural networks. In: 2016 IEEE International Conference on Acoustics, Speech and Signal Processing (ICASSP), pp. 2657–2661. IEEE (2016)

14. Liaw, A., Wiener, M., et al.: Classification and regression by randomforest. R News **2**(3), 18–22 (2002)
15. Makridakis, S.: A survey of time series. Int. Stat. Rev./Revue Internationale de Statistique **44**, 29–70 (1976)
16. Makridakis, S., Hibon, M.: The m3-competition: results, conclusions and implications. Int. J. Forecast. **16**(4), 451–476 (2000)
17. Min, W., Wynter, L.: Real-time road traffic prediction with spatio-temporal correlations. Transp. Res. Part C: Emerg. Technol. **19**(4), 606–616 (2011)
18. Moreira-Matias, L., Gama, J., Ferreira, M., Damas, L.: A predictive model for the passenger demand on a taxi network. In: 2012 15th International IEEE Conference on Intelligent Transportation Systems, pp. 1014–1019. IEEE (2012)
19. Moreira-Matias, L., Gama, J., Ferreira, M., Mendes-Moreira, J., Damas, L.: Predicting taxi-passenger demand using streaming data. IEEE Trans. Intell. Transp. Syst. **14**(3), 1393–1402 (2013)
20. Ntoutsi, E., Mitsou, N., Marketos, G.: Traffic mining in a road-network: how does the traffic flow? Int. J. Bus. Intell. Data Min. **3**(1), 82–98 (2008)
21. Sanders, N.R.: Measuring forecast accuracy: some practical suggestions. Prod. Inventory Manage. J. **38**(1), 43 (1997)
22. Seltman, H.: Simple linear regression. Chapter **9**, 217–240 (2015)
23. Srivastava, N., Hinton, G., Krizhevsky, A., Sutskever, I., Salakhutdinov, R.: Dropout: a simple way to prevent neural networks from overfitting. J. Mach. Learn. Res. **15**(1), 1929–1958 (2014)
24. Tong, Y., et al.: The simpler the better: a unified approach to predicting original taxi demands based on large-scale online platforms. In: Proceedings of the 23rd ACM SIGKDD International Conference on Knowledge Discovery and Data Mining, pp. 1653–1662. ACM (2017)
25. Unnikrishnan, V., et al.: Entity-level stream classification: exploiting entity similarity to label the future observations referring to an entity. Int. J. Data Sci. Anal. 1–15 (2019). https://doi.org/10.1007/s41060-019-00177-1
26. Wong, S.C., Gatt, A., Stamatescu, V., McDonnell, M.D.: Understanding data augmentation for classification: when to warp? In: 2016 International Conference on Digital Image Computing: Techniques and Applications (DICTA), pp. 1–6. IEEE (2016)
27. Xu, J., Wong, S., Yang, H., Tong, C.O.: Modeling level of urban taxi services using neural network. J. Transp. Eng. **125**(3), 216–223 (1999)
28. Yao, H., et al.: Deep multi-view spatial-temporal network for taxi demand prediction. In: Thirty-Second AAAI Conference on Artificial Intelligence (2018)

Multi-channel Convolutional Neural Networks for Handling Multi-dimensional Semantic Trajectories and Predicting Future Semantic Locations

Antonios Karatzoglou(✉)

Robert Bosch GmbH, Chassis Systems Control, Advance Engineering Sector, Abstatt, Germany
antonios.karatzoglou@de.bosch.com

Abstract. Current location-aware systems rely increasingly on location prediction techniques in order to provide their services in a timely fashion. At the same time, it has been shown that the use of additional context information, that is, elevating the degree of semantic enrichment of movement data, can lead to significant better results both in analyzing as well as in modeling human trajectories and predicting upon them. In this work, we propose a Multi-Channel Convolutional Neural Network (CNN) based approach for capturing all the available context dimensions in our semantic trajectory dataset aiming at achieving a higher prediction accuracy compared to a vanilla locations-only Single-Channel CNN. Moreover, we investigate whether and to what degree time, activity, companionship and the user's emotional state have an impact on the predictive performance of our multi-dimensional CNN. We evaluate our model using a real-world dataset and compare it among others to a probabilistic Markov Chain model and a vanilla CNN at two semantic representation levels. It can be shown that especially for higher level representations, the present approach is able to outperform the baseline models achieving an overall higher accuracy and F1-Score.

Keywords: Semantic locations and trajectories · Location prediction · Purpose of visit · Human activities · Emotional states · Context awareness · Multi-channel convolutional neural networks

1 Introduction

The market of context-aware, and especially, location-aware computing and services (see *Location-Based Services (LBS)*) has gained enormously in importance over the past few decades. In their attempt to provide timely solutions to their users, LBS providers rely more and more on location prediction methods, a fact that additionally strengthened the demand for accurate location prediction algorithms.

K. Tserpes et al. (Eds.): MASTER 2019, LNAI 11889, pp. 117–132, 2020.
https://doi.org/10.1007/978-3-030-38081-6_9

Typical location prediction models are usually merely data driven and depend therefore heavily on the size and the quality of the available training datasets. However, recent research has shown that the use of additional semantic information can help overcome, at least to some extent, the aforementioned dependencies and can therefore lead to an overall better predictive performance (see Sect. 2). That is, in the case of modeling and learning human movement patterns, models are usually fed and trained with the users' plain GPS location trajectories. Further context information, such as the location type and the user's activity, may be however used to enrich these semantically and generate so-called *semantic trajectories* (see Sect. 3). This type of extensive input helps the model dive even deeper into the users's movement behaviour and can lead to more accurate predictions.

Common approaches used to model and predict human movement include probabilistic methods, such as Markov Chains [?] [9], Dynamic Bayes Networks [8], Hidden Markov Models [26,31] and Artificial Neural Networks (ANNs). In the latter case, recurrent neural network architectures (RNNs) have generally proved to perform above the average when it comes to learning sequences and for this reason these are commonly found in the location prediction domain as well. Especially memory-based neural network types, like the Long Short-Term Memory network (LSTM), are capable of achieving high prediction rates and tend to outperform the competition [16,28].

While recurrent network types are the preferred choice when it comes to modeling 1-dim movement patterns, recent work shows some promising results on the part of Convolutional Neural Networks (CNNs) [17,25], a model normally used in the 2-dim image classification and object recognition domain. It seems that the locally focused nature of the kernel-based convolution process enables the CNN model to successfully capture existing dependencies between current and future locations found in the data. The presented study builds upon this work and aims at investigating the use of a *multi-channel CNN* based approach with regard to modeling *multi-dimensional semantically enriched location data* and predicting the next semantic location of the user. In particular, our semantic trajectories consist of the following feature dimensions: *semantic location type, time, human activity, emotional state and companionship.* Moreover, this work further explores the impact of the degree of semantic enrichment, that is, whether and to what extent each of the aforementioned dimensions influences the predictive performance of our model. We evaluated our approach using a real-world dataset, which we collected from 21 users by conducting a 2-months long user study. In addition, we selected a 1. Order Markov Chain model and a vanilla CNN to be our baseline.

This paper is structured as follows. Section 2 provides a short overview over some of the most related work in the semantic trajectories and location prediction domain. Next, Sect. 3 describes the notion of *semantic trajectories* and *semantic locations* with respect to this work. Section 4 goes briefly through the theory behind Convolutional Neural Networks and discusses in detail the proposed approach, while Sect. 5 provides the respective evaluation outcomes. Finally, Sect. 6 summarizes the evaluation results and draws some final conclusions.

2 Related Work

There exist many different ways of viewing movement data. Within the scope of mining and analyzing movement patterns, Spaccapietra et al. introduced with [29] one of the first works that make the importance of viewing trajectories of moving objects in a conceptual manner clear. In their work, they highlighted the fact that describing certain aspects of the movement's context by adding semantic information into the available trajectories can significantly support the analysis of the respective movement patterns, as well as the querying process among them. Alvarez et al. came to the same conclusion as they suggested the use of a similar semantic enrichment model to generate *semantic trajectories* for the same reasons [1]. The added value of working on semantically enriched trajectory data in comparison to working on raw data with regard to mining patterns and supporting decision processes has been underpinned by Elragal et al. as well [7]. Bogorny et al.'s work also focuses in mining trajectory patterns and introduced in [2] a sophisticated model, which in contrast to former models is capable of handling complex queries over semantic trajectories, while providing different semantic granularities at the same time. Finally, Karatzoglou et al. showed in [20] that considering multiple context dimensions results in generating more accurate synthetic semantic trajectories.

Due to the aforementioned benefits that accrue from semantic enrichment, a number of location prediction papers have recently emerged presenting algorithms that rely on the notion of semantic trajectories. Ying et al. for instance were one of the first to build upon semantic trajectories generated from the users' raw GPS recordings in order to enhance their location prediction framework [32] with promising results. Some years later, they extended their model by taking, apart from geographic and semantic patterns, temporal patterns into account as well [33].

Karatzoglou et al.'s work explores a big variety of models with respect to modeling human semantic trajectories and predicting the user's next semantic location. In [12] and in [18] they evaluate a multi-dimensional Markov Chain model with respect to predicting among activity-enriched semantic trajectories and show that it is able to outperform Ying et al.'s framework in terms of accuracy. With regard to recall however, they could identify certain limitations on behalf of the model due to its adverse dependency on the small size and the sparsity of the available training dataset. They attempt to solve this issue by combining the probabilistic Markov Chain model with Matrix Factorization in [11], where they were able to raise the recall scores.

In [13,16,17,19], Karatzoglou et al. investigate the performance of Artificial Neural Networks using the probabilistic Markov model as baseline. In addition, they explore the role of the semantic granularity of the considered trajectories in the overall performance of the networks. They show that the higher the semantic level is, the better the modeling quality of the networks. While the findings in [16] comply with the results of related work showing that Long Short-Term Memory networks are generally able to outperform the vanilla Recurrent (RNN) and the Feed-Forward model, [19] indicates no great advantages towards the attention-based application of Sequence to Sequence learning (Seq2Seq) compared to the standard single-input-single-output LSTM model of [16], a fact that may primarily explained by the limited size of the training dataset. Yao et al. propose in [30] a similar to [16] LSTM-based recurrent approach for predicting next semantic locations using an additional embedding input layer, the benefits of which have been also recently shown by Gao et al. in [10]. Other than in [16] and following a similar direction to the approach proposed in the present paper, Yao et al. used beyond location and time the content of the checkins of the users to enrich the users' semantic trajectories, which describe in a way their activity that we're considering in this work as well (among other). However, in contrast to the Reality Mining dataset [6] used in Karatzoglou et al.'s work, they evaluate their approach on rather long-term dependencies using a Foursquare and a Twitter dataset. In [13], Karatzoglou et al. take a look at a gradient-free optimization method for finding the optimal hyperparameter set of a LSTM model based on an evolutionary algorithm. Their work provides some first preliminary results indicating among others the temporal efficiency on part of the genetic, population-based optimization method, provided the fact that sufficient computational power is available.

The most striking findings come rather from [17], where a Convolutional Neural Network based approach in combination with an embedding layer as its input is capable of achieving higher prediction scores than the FFNN, the RNN and the LSTM. To our knowledge this represents the only work that explores the use of CNNs with respect to modeling and predicting upon 1-dim semantic trajectories. The closest work to [17] would be the work of Lv et al. in [25], which evaluates the use of a CNN for modeling and predicting large-scale taxi trajectories. Other than in [17] and the present paper, Lv et al. work with raw GPS data without using any semantic information and map past trajectory data onto 2-dim images before feeding them into the CNN model, transforming in this way the trajectory modelling task into an image classification task. In the present paper, following the example of [17], we skip this kind of 1-dim to 2-dim intermediate transformation step and apply our CNN model on the 1-dim semantic trajectory as it is. As in [17], we build our approach upon similar CNN-based work on 1-dim data, work, that comes mostly from the Natural Language Processing (NLP) domain, such as the framework described in [4] and the multi-channel CNN model of [22].

3 Semantic Trajectories

The term *trajectory* refers to a sequence of consecutive location points traversed by a moving object within a certain time interval. Equation 1 describes a typical GPS trajectory with each location point being represented by a tuple containing its coordinates ($long_i$, lat_i) and the corresponding point of time t_i at which this was visited.

$$Traj_{GPS} = (long_1, lat_1, t_1), (long_2, lat_2, t_2), ..., (long_i, lat_i, t_I), \quad (1)$$

As already mentioned in Sect. 2, in order to better understand the moving behaviour of moving objects and create more accurate models, Spaccapietra et al. [29] and Alvares et al. [1] went beyond this kind of numerical sequences by adding a semantic view upon them and introduced the so-called *semantic trajectories*. Starting initially with the simple notion of "stops" and "moves", a (human) semantic trajectory can nowadays be defined generally as a sequence of semantically significant locations (*semantic locations*, e.g., "home", "burger joint", etc.) as follows:

$$Traj_{Sem} = (SemLoc_1, t_1), (SemLoc_2, t_2), \ldots, (SemLoc_i, t_i) \quad (2)$$

A significant location in this case is usually defined by a location within a certain radius (e.g., 200 m) where a user stays longer than a pre-defined temporal threshold, e.g. 20 min (see [?]). Some researchers add further thresholds, like the loss of the GPS signal due to entering into a building, the GPS recording stop [3] or the popularity, in order to extract the most significant common or public locations [32]. In this work, we evaluate our method using a dataset in which the users annotated their longest visits (>15 min) by themselves (see Sect. 5).

Depending on the number of the considered semantic features, a semantic trajectory can have multiple dimensions. Thus, we could say that the number of dimensions expresses the *degree of semantic enrichment* of the respective semantic trajectory. In this work, we follow the concept of the Location-Specific Cognitive Frames introduced by Karatzoglou et al. in [14,15] and we consider each stop at a semantic location to be a tuple encapsulating the current location type, the current time, the current activity, as well as the user's current emotional state and whether he is alone or not (companionship). Beyond that, locations can be described differently depending on the semantic representation level, e.g., "restaurant" → "fast food restaurant" → "burger joint". In this work, we evaluate the modeling performance of a CNN at two different semantic levels. That is, we evaluate two different models, one that is trained for handling and predicting low level trajectories and one for handling higher level ones.

4 Multi-channel Convolutional Neural Networks on Semantic Trajectories

This section consists of two parts. The first part gives a brief insight into the theory behind Convolutional Neural Networks and goes briefly through some of the most common CNN steps and layers using the example of image classification. Then, the second and last part describes in detail the architecture of the multi-channel CNN model proposed in this paper for handling multi-dimensional semantic trajectories.

Convolutional Neural Networks (CNN) constitute the state of the art choice in the image classification and object recognition domain [24]. However, this doesn't mean that it is the only domain in which we can apply them expecting reasonable results as we saw in Sect. 2 and can also be seen in [23]. Figure 1 illustrates a typical CNN pipeline used for classifying images.

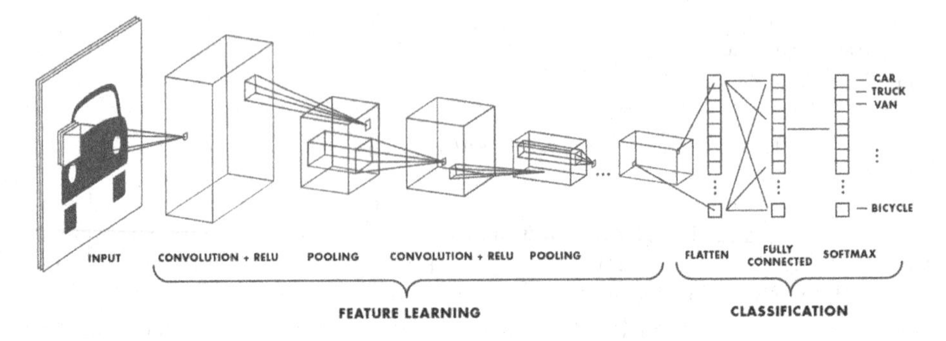

Fig. 1. Typical CNN architecture for the image classification task (source: [27]).

A typical CNN consists of many different layers starting usually with the (first) convolutional layer. This layer is responsible for convolving the input image and generating the so-called *feature maps*. This is done by sliding a group of small-sized *filters* (also called *kernels*) with each containing a certain number of learnable weights over the input image and performing element-wise multiplication at each possible position. The generated feature map from each kernel is a new layer and contains the findings of the particular kernel in the input image, ideally with respect to a specific and distinguished single feature. The number of kernels defines the number of the generated feature maps (so-called *depth* of the convolutional layer) and represents a CNN hyperparameter which needs to be selected appropriately based on the available data and task. In the next step, this resulting group of layers undergoes a so-called *pooling* process. Pooling refers to a downsampling operation, in which sets of elements in the feature maps are combined and reduced to a single value based on some criterion (e.g., take the maximum value: *max pooling*) or on some type of calculation (e.g., take the average over all values: *average pooling*). The two aforementioned layers can be repeated multiple times using different kernels of different size and depth. This supports the successive extraction of higher level features and represents one of the strengths of CNNs. Finally, the last pooled layer can be *flattened* into a single vector containing all its weights and connected to a *fully connected layer*, which is further connected to the output layer that contains a field for every possible class and provides us with the classification estimation for the given input.

The multi-channel approach introduced in this paper builds upon the aforementioned typical CNN architecture and extends it by adding a further *embedding layer* into the model and by *raising the number of channels* matching

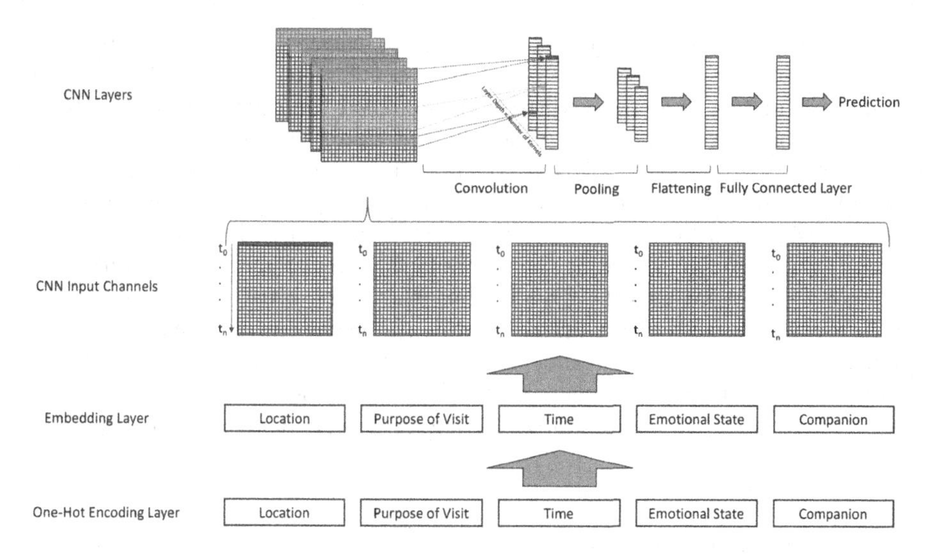

Fig. 2. Multi-channel CNN for modeling multi-dimensional semantic trajectories.

the degree of semantic enrichment of our data (see Sect. 3). Figure 2 illustrates the architecture of our approach. Our framework takes as input a part of a semantic trajectory, which is in our case a sequence of tuples in the form of $(locationtype, purposeofvisit, time, emotionalstate, companionship)$ according to a predefined temporal horizon t_n that determines how far backwards in the movement history of the user should the model consider for providing an estimation about her next semantic location. In a first step, every single feature type is encoded as a one-hot vector. The additional embedding layer between the one-hot encoded input and the convolutional layer maps the sparse asymmetrical one-hot encoded binary vectors into dense vector representations in a continuous vector space. This fact contributes to a more efficient training and helps improving the prediction accuracy while keeping the model consistent at the same time. The number of dimensions of the vector space is selected based on the properties of the available data, e.g., the number of unique classes of the corresponding feature. In our case, each semantic feature is encoded separately, and therefore the generated vectors may have different number of dimensions.

Raising the number of channels according to the semantic enrichment degree of our trajectories represents an intuitive way of viewing upon them. Each channel handles solely the corresponding semantic dimension. For example the first channel is responsible for the location type, the second channel for the purpose of visiting that location (activity), the third one for covering temporal information and so on. At the end, all channels are merged into a single representation, flattened and forwarded to the output layer in order to provide a final prediction about the next semantic location of the user. It should be noted here that the kernels' depth should match the channel dimension (5 in our case).

Our CNN has one convolutional, one pooling, a flattening, a fully connected and a Softmax output layer. A deeper architecture, that is, adding more layers led in most of the cases to overfitting and reduced the overall performance of our model due to the small size of our dataset compared to the higher parameter number. Other than the CNN model in Fig. 1, our model executes a 1-dim convolution operation instead of the typical 2-dim operation conducted in the image classification task. Each kernel convolves each semantic dimension in one direction only, namely according the chronological order found in the input tuple sequence which is fed into the model. Thus, the width of each kernel covers the whole row of the CNN input matrices while its height can vary, constituting a further hyperparameter of our model. A higher height indicates a kernel, able to observe a higher number of consecutive locations at the same time, a fact that can be useful when aiming at capturing long-term dependencies in our data, and vice versa.

5 Evaluation

In this section, we evaluate a multi-user version of our approach, which is trained on location data coming from multiple users. For this purpose, we first concatenated the trajectories of all users to a single trajectory, ordered by the user ID. Then, we randomly split the resulting trajectory into a training and a test dataset with a ratio of 80% to 20% while maintaining the user order at the same time (i.e., without breaking a user's trajectory into 2 parts). All in all, we randomly split the data 3 times and the findings in this section refer to the average over these 3 runs.

In order to evaluate our approach, we conducted a 8-week long user study tracking 21 users via a tracking and annotation app. The participants of the study were asked to semantically label each significant stop (location type) during their movement, as well as to note the purpose of visiting the certain location (e.g., activity), their companion (if any) and their emotional state by selecting among the following states: happy, hungry, neutral, sleepy, energetic, frustrated, stressed, bored, adventurous, ill, sad, angry and shocked. At the end of the study we end up with approximately 1400 annotated locations covering around 70 unique location types, 53 unique activities, and all 13 emotional states. A thorough description of the user study can be found in [21].

In order to take time into account, we defined $24 \times 7 = 168$ hourly slots, which similar to the other input signals were one-hot encoded first and transformed into an embedding vector in a next step. However, our evaluation results showed that taking time into account had a severe negative impact on the prediction outcome of our model. We saw a similar behaviour in the work of Karatzoglou et al. in [16] and in [18]. This can be mainly attributed to the small size of our dataset which makes it extremely hard for the model to find temporal patterns in this 168-slot temporal granularity. The use of wider time slots, e.g., the use of just daily slots, couldn't yield significantly higher scores either, due to the fact that our 8-week long evaluation dataset contains solely 8 recordings from each day, that is, there exist solely 8 unique Mondays, 8 unique Tuesdays, etc. For this reason, and due to space reasons, this evaluation section neglects to refer thoroughly to the individual results with respect to time. In addition, our users provided very little information regarding their type of companionship (e.g., relative, friend, etc.). Solely the fact whether a user was alone or not can be reliably extracted from our dataset. Therefore, instead of handling the companionship in a separate channel, we integrated the particular information into the emotional state one-hot vector by extending it to a further dimension ('0', when the user is alone and '1' when he is not). Finally and as already mentioned previously in this work, we evaluated our approach at two semantic representation levels, which will be referred to as *low* and *high* level, with the latter being more abstract and subsuming the first one. In order to generate these two layers, we built a corresponding location taxonomy based on the Foursquare venue categorization[1]. Lastly, a Grid Search helped us to determine the following optimal hyperparameter configuration for our model listed in Table 1.

Table 1. Optimal hyperparameter set determined via Grid Search.

Kernel size	Number of kernels	Embedding dimension	Dropout probability	Batch size	Learning rate	Number of epochs	Sequence length	Pooling size
6	64	100	0.6	16	0.001	100	10	2

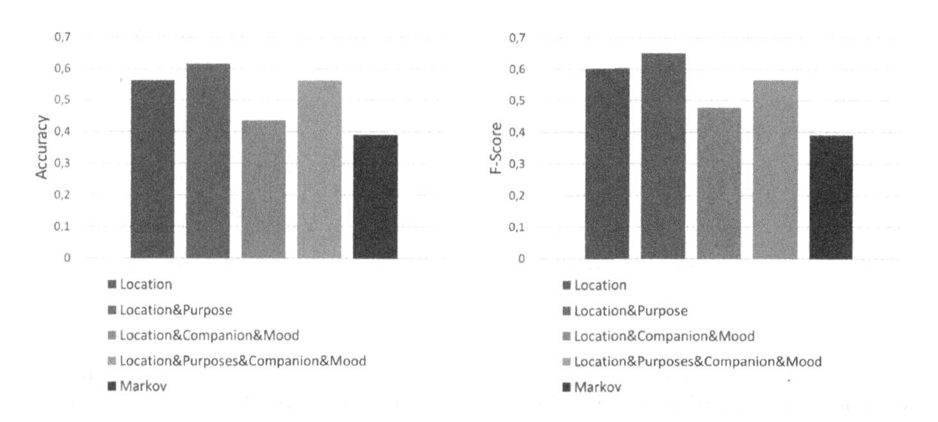

Fig. 3. Accuracy and F1-Scores at the higher semantic representation level.

Figure 3 compares the result from 5 different models at the higher representation level, a standard 1-channel CNN (*Location*) that takes just the current semantic location as input, a 2-channel CNN that considers the location

[1] https://developer.foursquare.com/docs/resources/categories.

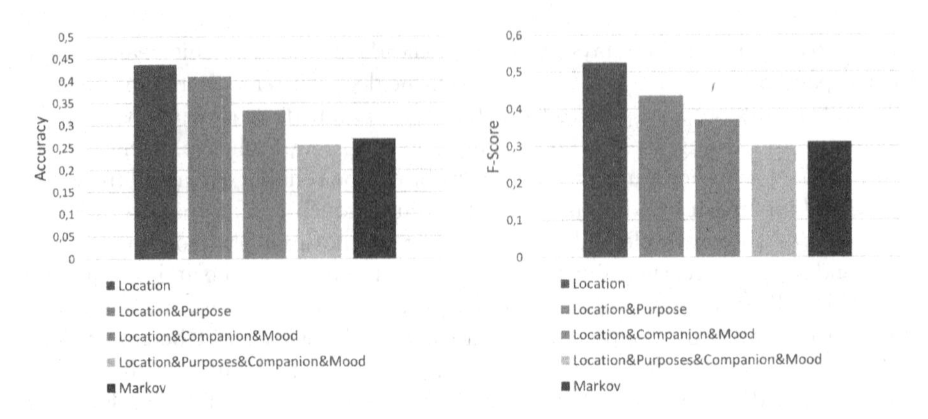

Fig. 4. Accuracy and F1-Scores at the lower semantic representation level.

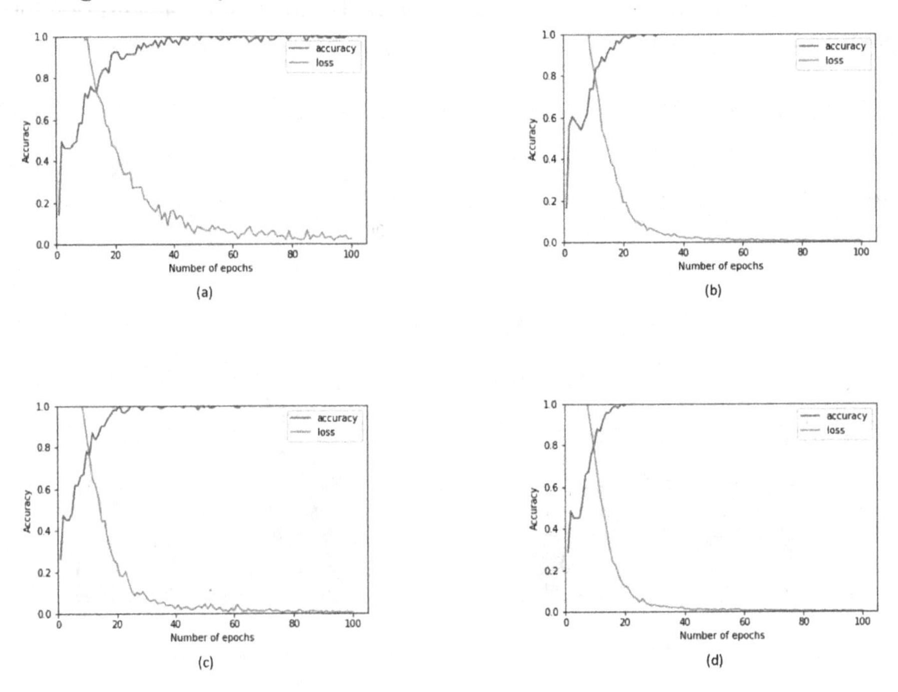

Fig. 5. Training accuracy and loss curves at the higher representation level. (a): Location, (b): Location&Companion&Mood, (c): Location&Purposes, (d): Location&Purposes&Companion&Mood.

type and the purpose of visit (*Location&Purpose*), a 2-channel CNN that considers the location type and the emotional state as well as the companionship status of the user (*Location&Companion&Mood*), a 3-channel CNN that takes location type, purpose of visit, emotional state and companionship (*Location&Purposes&Companion&Mood*) into account, and a probabilistic Markov

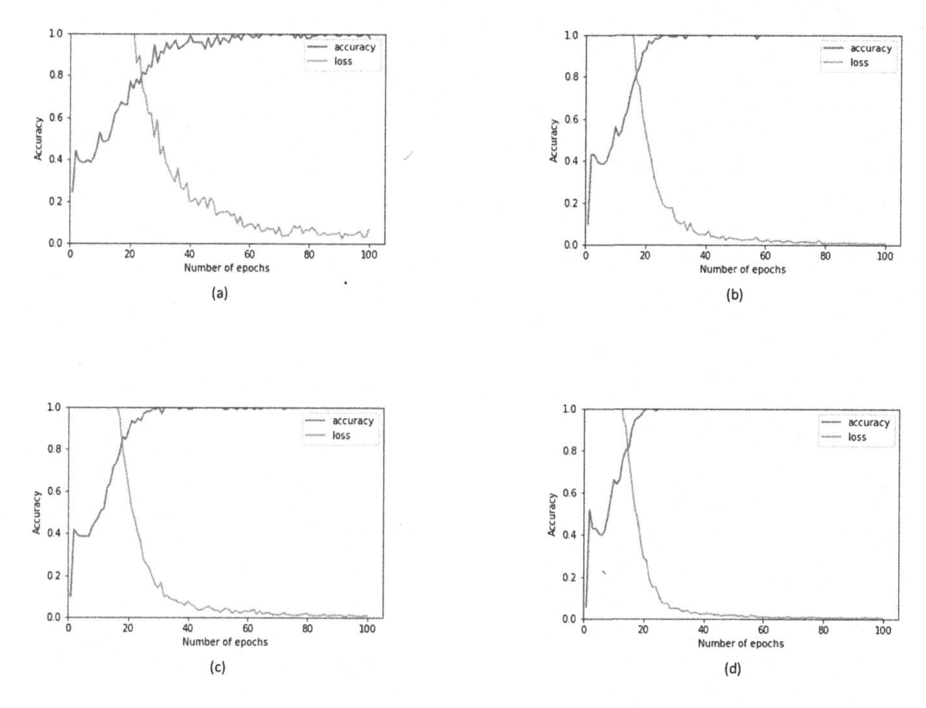

Fig. 6. Training accuracy and loss curves at the lower representation level. (a): Location, (b): Location&Companion&Mood, (c): Location&Purposes, (d): Location&Purposes&Companion&Mood.

Chain model of 1. order that serves as our reference. It can be seen that all the CNN-based approaches are able to outperform the Markov model both in terms of accuracy and F1-Score. What also stands out in the same figure is that the 2-channel CNN approach that considers the activity of the user (purpose of visit) can clearly outperform the competition. However, this doesn't hold for the other 2-channel CNN model. On the contrary, it seems that taking the user's emotional state into account affects negatively the predictive performance. Apparently, our model wasn't able to establish an association between the users' movement behaviour and their mood, a fact that could be partly attributed once again to the small size of our dataset. The more "sophisticated" 3-channel CNN achieves a similar accuracy to the standard CNN, but a lower F1-Score and therefore can't really compete with the *Locations&Purposes* model. Its results are likely to be related to the aforementioned negative impact of the emotional state when this is taken explicitly into account.

Figure 4 presents the results for the lower semantic representation level. It is apparent that all models perform worse than in the higher level shown in Fig. 3. This can be mainly attributed to the fact that the lower semantic representation carries a higher number of unique classes to predict, which makes the learning process of the models much harder. At the same time, another possible

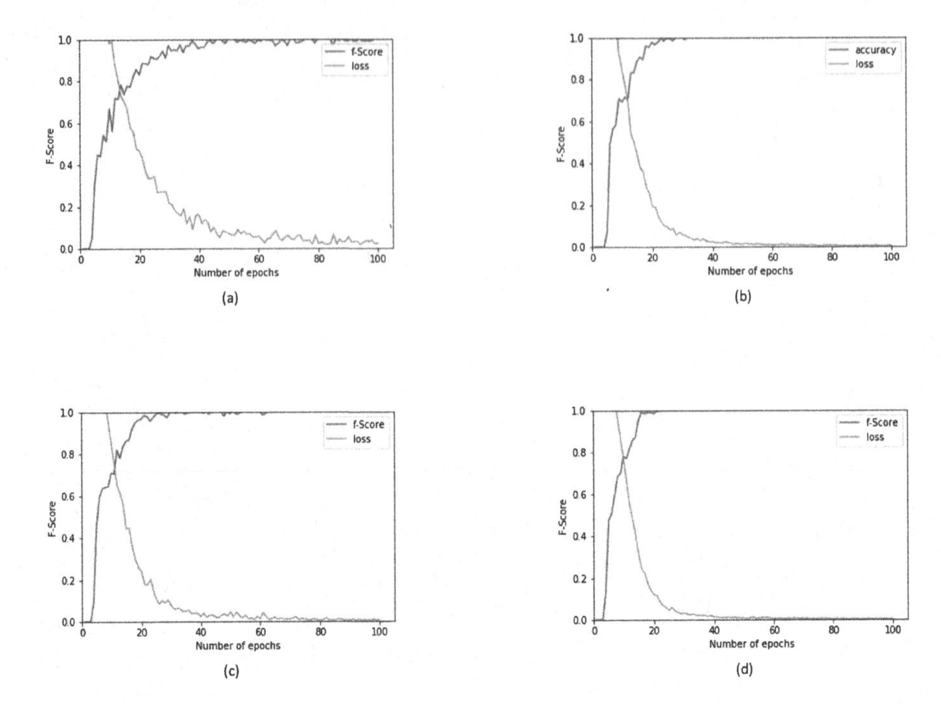

Fig. 7. Training F1-Score and loss curves at the higher representation level. (a): Location, (b): Location&Companion&Mood, (c): Location&Purposes, (d): Location&Purposes&Companion&Mood.

explanation for this might be the fact that human movement shows stronger regularities at rather higher levels, e.g., a user may often visit a food location after going to gym, regardless whether this location is an Italian or a Greek restaurant, a pizza house or a burger joint. Similar to the high level case, the CNN models outperform in most of the cases the probabilistic baseline. However, this time, other than at the higher level, it seems that the additional channels result in a deterioration of our prediction models. The more channels, the worse the predictive behaviour seems to become. In general, due to the small size of our dataset and its imbalance, all of our models had to deal with massive overfitting issues. Adding a dropout layer while making our model simpler by reducing the size of our layers could improve significantly our models, but only to a certain extent.

Figures 5, 6, 7 and 8 illustrate the training behaviour of our 4 CNN models. We can see that the greater the number of channels and thus, the greater the semantic enrichment degree of the trajectory, the faster and smoother the training of the CNN model becomes. Taking additional context dimensions into account seems to contribute to shorter convergence times and results in a more efficient training. The 3-channel CNN is characterized by the shortest convergence, while the vanilla 1-channel CNN straggles with the loss reduction along the whole training process to the 100th epoch. The benefits of the multiple chan-

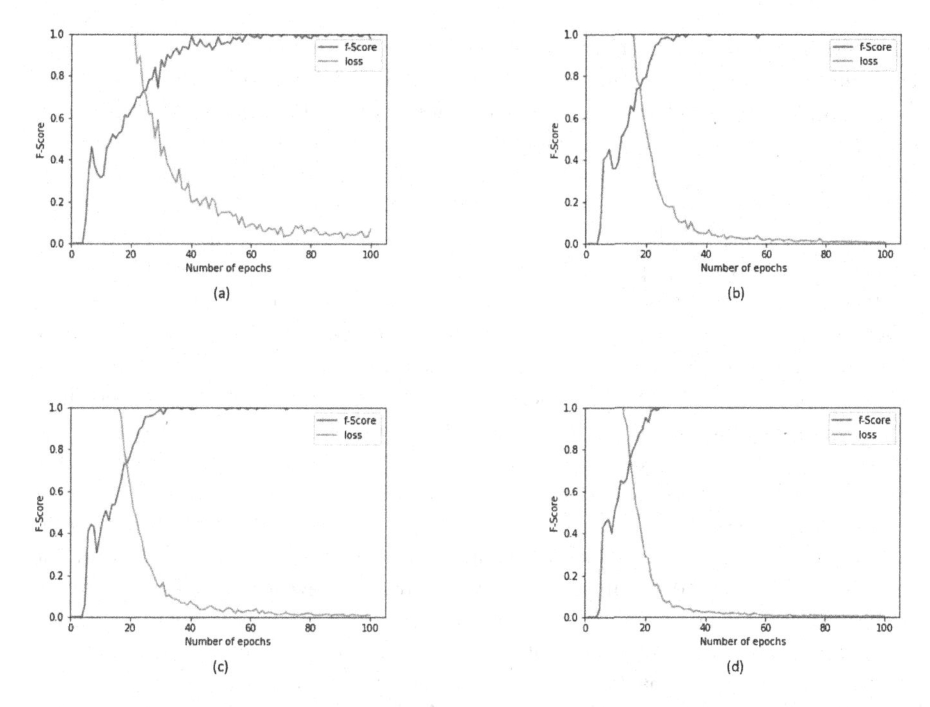

Fig. 8. Training F1-Score and loss curves at the lower representation level. (a): Location, (b): Location&Companion&Mood, (c): Location&Purposes, (d): Location&Purposes&Companion&Mood.

nel approach can be more clearly seen during the harder learning task, namely at the lower semantic representation level (see Fig. 6). However, on the other hand, this comes with a certain overfitting effect, as mentioned previously, that grows with the number of the CNN input channels.

The models presented in this work use 1-dim fixed-sized kernels of size 6. This number was determined by applying a Grid Search. The problem when using fixed-sized kernels is that these are able to capture only data dependencies of a certain length. Varied-sized kernels as in the work of Kim et al. in [22] could help overcome this issue and capture the individual properties of each semantic dimension in our data.

6 Conclusion

In this work, we explore the performance of a Multi-Channel Convolutional Neural Network (CNN) based approach with respect to its capability of modeling semantic trajectories at different semantic representation levels and predicting the next semantic location of a user. Moreover, we investigate whether and to what extend the degree of semantic enrichment, that is, the number of the context feature dimensions considered in the semantic trajectories, affects the predictive performance of our model. We considered 5 different semantic enrichment dimensions for our trajectories, the location type, the purpose of visit (e.g., activity), the time, the user's mental and emotional state, and his companionship. We evaluated our model using a 8-week long real-world dataset from 21 users and compared it to a vanilla Single-Channel CNN and a probabilistic Markov Chain model that served among other as our baseline. We could show that raising the semantic enrichment degree of our trajectory data while increasing the corresponding number of channels at the same time can indeed lead to an improvement in terms of prediction accuracy and F1-Score. This could be particularly seen when we attempted to model and predict upon semantic trajectories at a higher representation level. Furthermore, the results of this work indicate a strong correlation between the degree of semantic enrichment, the number of CNN channels and the training behaviour, with our multi-channel based approach being characterized by generally much smoother and faster converging learning curves. However, our evaluation also identified some limitations leaning mostly on certain overfitting effects, which could be mainly attributed to data-specific properties, such as the small size of our dataset and its imbalance. This is also the reason why a certain uncertainty about the generalizability and the representativity of the findings in this work arises. Nevertheless, the presented study still establishes a solid basis for further work and investigations. In our future work, we plan to further explore the use of CNNs in the semantic location prediction scenario. In particular, we plan to investigate the use of varied-sized kernels and depthwise separable convolution layers aiming at improving both the predictive performance as well as the computational efficiency of our model. Furthermore, we would like to experiment with further types of context information, such as the personality of the user, the weather and the transportation mode; features, that have led to promising results in existing studies. However, the gathering of context information and especially of personal information has become increasingly difficult in recent years, among others, due to stricter data privacy regulations. One solution for overcoming this issue would be to rely on privacy preserving methods such as the semantic obfuscation techniques of [5].

References

1. Alvares, L.O., et al.: A model for enriching trajectories with semantic geographical information. In: Proceedings of the 15th Annual ACM International Symposium on Advances in Geographic Information Systems, p. 22. ACM (2007)
2. Bogorny, V., et al.: Constant-a conceptual data model for semantic trajectories of moving objects. Trans. GIS **18**(1), 66–88 (2014)
3. Cao, X., et al.: Mining significant semantic locations from GPS data. Proc. VLDB Endow. **3**(1–2), 1009–1020 (2010)
4. Collobert, R., et al.: Natural language processing (almost) from scratch. J. Mach. Learn. Res. **12**, 2493–2537 (2011)
5. Damiani, M.L., Bertino, E., Silvestri, C.: Protecting location privacy through semantics-aware obfuscation techniques. In: Karabulut, Y., Mitchell, J., Herrmann, P., Jensen, C.D. (eds.) IFIPTM 2008. ITIFIP, vol. 263, pp. 231–245. Springer, Boston, MA (2008). https://doi.org/10.1007/978-0-387-09428-1_15

6. Eagle, N., Pentland, A.S.: Reality mining: sensing complex social systems. Pers. Ubiquitous Comput. **10**(4), 255–268 (2006)
7. Elragal, A., El-Gendy, N.: Trajectory data mining: integrating semantics. J. Enterp. Inf. Manag. **26**(5), 516–535 (2013)
8. Etter, V., et al.: Been there, done that: what your mobility traces reveal about your behavior (2012)
9. Gambs, S., et al.: Next place prediction using mobility Markov chains. In: Proceedings of the First Workshop on Measurement, Privacy, and Mobility, MPM 2012, pp. 3:1–3:6. ACM, New York (2012)
10. Gao, Q., et al.: Identifying human mobility via trajectory embeddings. In: Proceedings of the 26th International Joint Conference on Artificial Intelligence, pp. 1689–1695. AAAI Press (2017)
11. Karatzoglou, A., et al.: Matrix factorization on semantic trajectories for predicting future semantic locations. In: 2017 IEEE 13th International Conference on Wireless and Mobile Computing, Networking and Communications (WiMob), pp. 1–7, October 2017
12. Karatzoglou, A., et al.: Purpose-of-visit-driven semantic similarity analysis on semantic trajectories for enhancing the future location prediction. In: 2018 IEEE International Conference on Pervasive Computing and Communications Workshops (PerCom Workshops), pp. 100–106, March 2018
13. Karatzoglou, A.: Evolutionary optimization on artificial neural networks for predicting the user's future semantic location. In: Macintyre, J., Iliadis, L., Maglogiannis, I., Jayne, C. (eds.) EANN 2019. CCIS, vol. 1000, pp. 379–390. Springer, Cham (2019). https://doi.org/10.1007/978-3-030-20257-6_32
14. Karatzoglou, A., Beigl, M.: Applying situation-person-driven semantic similarity on location-specific cognitive frames for improving the location prediction. In: 8th International Conference on Knowledge Engineering and Semantic Web (KESW) (2017)
15. Karatzoglou, A., Beigl, M.: Enhancing the affective sensitivity of location based services using situation-person-dependent semantic similarity. In: Proceedings of the Eleventh International Conference on Mobile Ubiquitous Computing, Systems, Services and Technologies, UBICOMM 2017, pp. 95–100 (2017)
16. Karatzoglou, A., Sentürk, H., Jablonski, A., Beigl, M.: Applying artificial neural networks on two-layer semantic trajectories for predicting the next semantic location. In: Lintas, A., Rovetta, S., Verschure, P.F.M.J., Villa, A.E.P. (eds.) ICANN 2017. LNCS, vol. 10614, pp. 233–241. Springer, Cham (2017). https://doi.org/10.1007/978-3-319-68612-7_27
17. Karatzoglou, A., Schnell, N., Beigl, M.: A convolutional neural network approach for modeling semantic trajectories and predicting future locations. In: Kůrková, V., Manolopoulos, Y., Hammer, B., Iliadis, L., Maglogiannis, I. (eds.) ICANN 2018. LNCS, vol. 11139, pp. 61–72. Springer, Cham (2018). https://doi.org/10.1007/978-3-030-01418-6_7
18. Karatzoglou, A., et al.: Semantic-enhanced multi-dimensional markov chains on semantic trajectories for predicting future locations. Sensors **18**(10), 3582 (2018)
19. Karatzoglou, A., et al.: A Seq2Seq learning approach for modeling semantic trajectories and predicting the next location. In: Proceedings of the 26th ACM SIGSPATIAL International Conference on Advances in Geographic Information Systems, SIGSPATIAL 2018, pp. 528–531. ACM, New York (2018)

20. Karatzoglou, A., et al.: Towards an affective semantic trajectory generator (ASTG). In: 14th IEEE International Conference on Wireless and Mobile Computing, Networking and Communications, WiMob 2018, Limassol, Cyprus, 15–17 October 2018, pp. 1–10 (2018)
21. Karatzoglou, A., et al.: Sentient destination prediction (under review). User Modeling and User-Adapted Interaction (UMUAI) (2019)
22. Kim, Y.: Convolutional neural networks for sentence classification. arXiv preprint arXiv:1408.5882 (2014)
23. LeCun, Y., et al.: Convolutional networks for images, speech, and time series. In: The Handbook of Brain Theory and Neural Networks, vol. 3361, p. 10 (1995)
24. LeCun, Y., et al.: Convolutional networks and applications in vision. In: Proceedings of 2010 IEEE International Symposium on Circuits and Systems (ISCAS), pp. 253–256. IEEE (2010)
25. Lv, J., et al.: T-CONV: a convolutional neural network for multi-scale taxi trajectory prediction. arXiv preprint arXiv:1611.07635 (2016)
26. Mathew, W., et al.: Predicting future locations with hidden Markov models. In: Proceedings of the 2012 ACM Conference on Ubiquitous Computing, UbiComp 2012, pp. 911–918. ACM, New York (2012)
27. Mathworks: Convolutional neural network (2018). https://www.mathworks.com/discovery/convolutional-neural-network.html. Accessed 19 Feb 2018
28. Song, X., et al.: DeepTransport: prediction and simulation of human mobility and transportation mode at a citywide level. IJCAI **16**, 2618–2624 (2016)
29. Spaccapietra, S., et al.: A conceptual view on trajectories. Data Knowl. Eng. **65**(1), 126–146 (2008)
30. Yao, D., et al.: SERM: a recurrent model for next location prediction in semantic trajectories. In: Proceedings of the 2017 ACM on Conference on Information and Knowledge Management, CIKM 2017, pp. 2411–2414. ACM, New York (2017)
31. Ye, J., et al.: What's Your Next Move: User Activity Prediction in Location-based Social Networks, pp. 171–179 (2013)
32. Ying, J.J.C., et al.: Semantic trajectory mining for location prediction. In: Proceedings of the 19th ACM SIGSPATIAL International Conference on Advances in Geographic Information Systems, pp. 34–43. ACM (2011)
33. Ying, J.J.C., et al.: Mining geographic-temporal-semantic patterns in trajectories for location prediction. ACM Trans. Intell. Syst. Technol. **5**(1), 2:1–2:33 (2014)

Author Index

Printed in the United States
By Bookmasters